REPRESENTATIONS OF THE ROTATION AND LORENTZ GROUPS

LECTURE NOTES
IN PURE AND APPLIED MATHEMATICS

1. *N. Jacobson*, Exceptional Lie Algebras
2. *L.-Å. Lindahl* and *F. Poulsen*, Thin Sets in Harmonic Analysis
3. *I. Satake*, Classification Theory of Semi-Simple Algebraic Groups
4. *F. Hirzebruch, W. D. Neumann,* and *S. S. Koh,* Differentiable Manifolds and Quadratic Forms
5. *I. Chavel*, Riemannian Symmetric Spaces of Rank One
6. *R. B. Burckel*, Characterization of C(X) among Its Subalgebras
7. *B. R. McDonald, A. R. Magid,* and *K. C. Smith,* Ring Theory: Proceedings of the Oklahoma Conference
8. *Yum-Tong Siu*, Techniques of Extension of Analytic Objects
9. *S. R. Caradus, W. E. Pfaffenberger, and Bertram Yood,* Calkin Algebras and Algebras of Operators on Banach Spaces
10. *Emilio O. Roxin, Pan-Tai Liu,* and *Robert L. Sternberg,* Differential Games and Control Theory
11. *Morris Orzech* and *Charles Small*, The Brauer Group of Commutative Rings
12. *S. Thomeier*, Topology and its Applications
13. *Jorge M. López* and *Kenneth A. Ross*, Sidon Sets
14. *W. W. Comfort* and *S. Negrepontis*, Continuous Pseudometrics
15. *Kelly McKennon* and *Jack M. Robertson*, Locally Convex Spaces
16. *M. Carmeli* and *S. Malin*, Representations of the Rotation and Lorentz Groups: An Introduction

Other volumes in preparation

REPRESENTATIONS OF THE ROTATION AND LORENTZ GROUPS

AN INTRODUCTION

M. Carmeli and S. Malin

Department of Physics
Ben Gurion University
Beer Sheva, Israel

Department of Physics and Astronomy
Colgate University
Hamilton, New York

MARCEL DEKKER, INC., New York and Basel

MARCEL DEKKER, INC.

270 Madison Avenue, New York, New York   10016

LIBRARY OF CONGRESS CATALOG CARD NUMBER: 76-3337

ISBN: 0-8247-6449-8

Current printing (last digit):
10 9 8 7 6 5 4 3 2

ACKNOWLEDGMENTS

Part of the research reported in this book was supported by the Israel National Academy of Sciences, Grant No. 19(B)P, and by the Colgate Research Council, U.S.A.

PRINTED IN THE UNITED STATES OF AMERICA

To the Memory

of

GIULIO RACAH

Teacher and Educator

PREFACE

The theory of groups has become one of the most useful mathematical theories in physics in recent years. In particular, the rotation and Lorentz groups have extensively been used in quantum mechanics and quantum field theory. Although there are many excellent books on both the theory of representations of the rotation group and that of the Lorentz group, all of them are usually intended for graduate courses and for researchers in the field.

The text of the present book is based on a one semester course given at the Ben Gurion University of the Negev to the last year under-graduate students of mathematics and physics by one of us (M.C.) in 1972-73. The same lectures were redelivered at the Solid State Research Laboratory, Aerospace Research Laboratories, Wright-Patterson Air Force Base, Ohio in summer 1973 to the Theoretical Solid State Research Group. The text was subsequently supplemented by the Appendix by S.M. so as to give the reader more details on the important case of spinor representations of the Lorentz group. As is well known these representations are most widely used in both theories of quantum mechanics and quantum fields, and in the theory of general relativity.

The present book is written as a text for undergraduate students of mathematics and natural sciences. It could also be used as a supplementary text for graduate students and for researchers who are

v

interested in using the theory without going in depth into the theory of representation. Although the monograph is written accurately from the point of view of mathematics, its approach is very elementary, and one semester will usually be enough for it. It has eleven chapters and an appendix. It includes all finite-dimensional representations of both the rotation and the Lorentz groups. It does not include, however, any infinite-dimensional representations which have just started to enter into the picture of the theory of elementary particle physics.

Chapter I is a brief review on group theory. It includes elementary concepts such as group, subgroup, normal subgroup, factor group, isomorphism and homomorphism. The basic concepts of the theory of finite-dimensional representations is then brought in Chapter II. The rotation group is subsequently introduced, and pure rotations are described in terms of the familiar Euler angles. This is done in Chapter III. In Chapter IV the group SU_2 is introduced and its relation to the group of pure rotations is pointed out. The important concept of invariant integration over the rotation group and the group SU_2 is introduced at this stage. This is done in Chapter V. Representations of the rotation group are extensively discussed, and the Wigner matrix elements $D^j_{mm'}$ are derived, in Chapters VI and VII. The discussion on the rotation group is concluded in Chapter VIII by finding out the angular momentum operators.

Unlike known texts on representations of the rotation group, in which the Euler angles are used to parametrize the representations, in the present text use is made of other angles defining the rotations. These are the magnitude of the rotation and the two spherical angles of the direction of the rotation.

The discussion on the Lorentz group starts in Chapter IX where orthochronous Lorentz transformations are discussed. A preliminary discussion on the finite-dimensional representations of the Lorentz group is given, and the spinor representations are subsequently outlined. The appendix then concludes the monograph with a somewhat detailed discussion on the spinor representations of the Lorentz group.

Finally, we have not prepared an extended bibliography of all the significant original papers underlying many of the developments recorded in this monograph. Among the following excellent books or monographs is to be found the remedy for one or more of these deficiencies:

(1) B.L. van der Waerden, *Modern Algebra*, Fredric Ungar
     Publishing Co., New York, 1953.

(2) L. Pontrjagin, *Topological Groups*, Princeton University
     Press, Princeton, New Jersey, 1946.

(3) E.P. Wigner, *Group Theory and its Application to the Quantum
     Mechanics of Atomic Spectra*, Academic Press, New York, 1959.

(4) M.A. Naimark, *Linear Representations of the Lorentz Group*,
     Pergamon Press, New York, 1964.

(5) I.M. Gelfand, M.L. Graev, and N.Y. Vilenkin, *Generalized
     Functions, Vol. 5: Integral Geometry and Representations
     Theory*, Academic Press, New York, 1966.

(6) I.M. Gelfand, R.A. Minlos, and Z. Ya. Shapiro, *Representations
     of the Rotation and Lorentz Groups and their Application*,
     Pergamon Press, New York, 1963.

(7)   W. Rühl, *The Lorentz Group and Harmonic Analysis*,
      W.A. Benjamin, Inc., New York, 1970.

We would like to thank many of our colleagues and students who helped us in shaping this monograph. In particular our thanks are due to Dr. Tom Collins, head of the Theoretical Solid State Research Group, Aerospace Research Laboratories, United States Air Force, whose hospitality made it possible for us to have a most fruitful interaction with his group, and to Dr. A.O. Barut, for the continuous interest and helpful advise we have received from him during our work in the theory of representations. Finally, our thanks are also due to Technology Incorporated, U.S.A., for the typing of the first part of the manuscript on behalf of the United States Air Force, and to Mrs. Y. Ahuvia for the retyping of the whole monograph.

M. Carmeli

S. Malin

CONTENTS

REPRESENTATIONS OF THE ROTATION AND LORENTZ GROUPS

CHAPTER I

REVIEW OF GROUP THEORY

Group theory is well covered in textbooks, among which are those of
Pontrjagin[1], van der Waerden[2], and Wigner[3]. This chapter is devoted to the
exposition of the fundamental concepts of the theory.

§ 1.1 Group and Subgroup

A set G of elements is called a *group* if the following *axioms* are
satisfied:

(1) There exists an operation in G which associates with each two
elements a, b of G a third element c of G. This operation is
called *multiplication*, and the element c is called the *product* of a
and b, c = ab;

(2) The multiplication is *associative*, i.e., if a, b, and c are
elements of G, then (ab)c = a(bc);

(3) G contains a *right identity*, i.e., an element e such that
ae = a for any element a of G; and

(4) For each element a of G there exists a *right inverse*
element, a^{-1}, such that aa^{-1} = e.

If the set G is finite the group G is called *finite* and the
number of elements of G is called its *order*. Otherwise the group G is

called *infinite*. If the product of any two elements a and b of G is commutative, ab = ba, the group is called *abelian*. In abelian groups the multiplicative notation ab is replaced by additive notation a+b and the group operation is called *addition*. The identity is called *zero* and denoted by 0, and the inverse of a is called the negative of a and denoted by -a.

Since the product of group elements is associative, one writes for (ab)c = a(bc) simply abc. The same holds for products of any number of elements. One can easily show that a right identity e is also left identity, ea = a, for any element a of G, and that a right inverse a^{-1} of a is also left inverse, $a^{-1} a = e$. Hence the inverse of a^{-1} is a. Moreover, it follows that both the identity and the inverse are unique. This allows the use of the notation of algebra such as $a^{m+1} = a^m a$, with $a^1 = a$, for any natural number m. Negative powers of a are introduced by $a^{-m} = (a^{-1})^m$, $a^0 = e$. Hence $a^p a^q = a^{p+q}$, and $(a^p)^q = a^{pq}$, where p and q are integers.

A set H of elements of a group G is called a *subgroup* of G if it is a group with the same law of multiplication which operates in G. A necessary and sufficient condition for a subset H of a group G to be a subgroup is that if H contains two elements a and b it must also contain, the element ab^{-1}.

§ 1.2 Normal Subgroup and Factor Group

Let G be a group and H a subgroup, and let a and b be two elements of G. One calls a and b *equivalent*,[4] a ∿ b, if ab^{-1} is an element of H. The group G is thus divided into classes of equivalent elements each called a *right coset* of H relative to G. It

follows that if A is a right coset of H and a is an element of A
then A = Ha.[5] Moreover, every subset of the form Hb is a right coset
and the subgroup H itself is one of the cosets. One can also introduce
left cosets of H, written in the form aH. They are obtained from an
equivalence relation such that $a \sim b$ if $a^{-1}b$ belongs to H.

A subgroup N of a group G is called an *invariant* or *normal
subgroup* of G if for every element n of N and a of G the
element $a^{-1}na$ belongs to N. It follows that a necessary and
sufficient condition for right and left cosets of a subgroup N to
coincide is that N be a normal subgroup.[6]

If N is a normal subgroup of a group G and A and B are two
cosets of N, A = Na, B = Nb, then AB is also a coset of N. The
multiplication of cosets thus defined satisfies the group axioms, and the
set of all cosets is called the *factor group* of G by the normal subgroup N and is denoted by G/N.

§ 1.3 <u>Isomorphism and Homomorphism</u>

A mapping f of a group G on another group G' is called
isomorphism if it (1) is one-to-one; and (2) preserves the
multiplication. G and G' are then called *isomorphic*. The inverse f^{-1}
of an isomorphism f is itself an isomorphism. An isomorphism of a
group onto itself is called *automorphism*. The aggregate of all
automorphisms of a group forms a group.

A mapping f of a group G on another G' is called *homomorphism*
if it preserves the operation of multiplication. The set N of all
elements of G which go over into the identity of G' under the
homomorphism is called the *kernel* of the homomorphism. If the kernel

coincides with the identity of G then the homomorphism is an
isomorphism. It follows that N is a normal subgroup of G, and G' is
isomorphic to G/N. The isomorphism between G' and G/N is called the
natural isomorphism. The mapping f of a group G on G/N defined by
associating with each element a of G the element $f(a) = A$ of G/N
containing a is a homomorphism, called the *natural homomorphism* of a
group on its factor group. If f is a homomorphism of a group G on
another group G' and H is a (normal) subgroup of G, then $f(H)$ is a
(normal) subgroup of G'. If f is a homomorphism of a group G on
another group G', and g is a homomorphism of G' on a third group
G", then the mapping gf is a homomorphism of G on G".

One finally notes that if f is a homomorphism of a group G on
part of another group G' then the set of all elements of G' which
are images of elements of G forms a subgroup of G'. Also, if
f^{-1} (H') is the set of all elements of G which go into $H' \subset G'$
under the homomorphism f, and if H' is a (normal) subgroup of the
group G', then f^{-1} (H') is also a (normal) subgroup of the group G.

NOTES AND REFERENCES

1. L. Pontrjagin, *Topological groups,* Princeton University Press,
    Princeton, New Jersey, 1946.

2. B.L. van der Waerden, *Modern Algebra,* Fredric Ungar Publishing Co.,
    New York, 1953.

3. E.P. Wigner, *Group Theory and its Application to the Quantum
    Mechanics of Atomic Spectra,* Academic Press, New York, 1959.

4. A *relation of equivalence* is said to be established in a set M if
   every two elements a, b of M are either equivalent, a ~ b, or not
   equivalent, a not ~ b. A relation of equivalence should be
   a) reflective: a ~ a; b) symmetric: if a ~ b then b ~ a; and c)
   transitive: if a ~ b and b ~ c then a ~ c. A relation of
   equivalence in M divides M into disjoint classes of equivalent
   elements.

5. If A and B are two subsets of a group G, one denotes by AB the
   subset of all elements of the form ab, where a ε A and b ε B.
   The subset A^{-1} denotes all elements a^{-1}, where a ε A. The
   subset A^{m+1} is defined by $A^{m+1} = A^m A$, where $A^1 = A$, the subset
   A^{-m} is defined by $A^{-m} = (A^{-1})^m$ for a natural number m, and the
   subset A^0 is the set containing the identity only.

6. Every group has at least two normal subgroups, the subgroup which
   includes only the identity, and the subgroup which coincides with
   the group itself. A group which has no normal subgroup except for
   these two subgroups is called *simple*.

CHAPTER II

BASIC CONCEPTS OF REPRESENTATIONS THEORY

In this chapter fundamental concepts of the theory of finite-dimensional representations are reviewed. For more details the reader is referred to the books of Wigner[1], Naimark[2], and of Gelfand, Graev, and Vilenkin[3] and others[4,5].

§ 2.1 Linear Operators

Let R be a linear space and x a vector in R. A function $T(x)$ is called an *operator* in R if for any vector x of R there corresponds a vector $y = T(x)$ of R. An operator T in R is then called *linear* if $T(x + y) = T(x) + T(y)$ and $T(\alpha x) = \alpha T(x)$, for any x, y of R and any complex number α. Addition of two operators A and B is defined in the space R by $(A + B)x = Ax + Bx$ for all vectors x of R. Similarily, multiplication by a number α and multiplication of operators A and B in the space R are defined by $(\alpha A)x = \alpha(Ax)$ and $(AB)x = A(Bx)$. If, furthermore, A and B are linear operators in R then $A + B$, αA, and AB are also linear operators in R.

Linear operators in a finite-dimensional space R can be represented as matrices by introducing a basis, let us say, $e_1, \ldots, e_n$ in R. Accordingly, if A is a linear operator in the space R, then Ae_k can be written as a linear combination of $e_1, \ldots, e_n$, or,

7

$$Ae_k = \sum_{j=1}^{n} A_{jk} e_j, \quad \text{for} \quad k = 1, \ldots, n.$$

A_{jk} are the elements of the matrix representing the operator A relative to the basis $e_1, \ldots, e_n$. One can show that the operator A is completely determined by its matrix A_{ij}. Furthermore, the operations of addition, multiplication by a number, and multiplication of operators correspond to the same operations of their matrices relative to a fixed basis.

§ 2.2 Finite-Dimensional Representations of a Group

Let G be a group and g an arbitrary element of G. A correspondence $g \to D_g$ of each element g of the group G to a linear operator D_g in a finite-dimensional space R is called a *representation* if: (1) $D_{g_1} D_{g_2} = D_{g_1 g_2}$ and (2) D_e is the unit element in R, where e is the identity element of G. The space R is called the *space of representation* and its dimension is called the *dimension of representation.*

Two finite-dimensional representations $g \to D_g$ and $g \to D'_g$ of the group G in two spaces R and R' having the same dimensions, respectively, are called *equivalent* if basis in the spaces R and R' can be chosen so that the matrices of the operators D_g and D'_g are identical. A subspace S of the space R is called *invariant* with respect to the representation $g \to D_g$ if for every vector x of S one finds that $D_g x$ is also a vector in S for all elements g of the group G. If there is no invariant subspaces in the space R with respect to the representation $g \to D_g$, except for the trivial cases of the null subspace and the whole space, the representation is then called *irreducible.* A representation $g \to D_g$ of a group G is called *continuous* if D_g is a continuous operator function on the group G[6].

We will consider here only continuous representations.

§ 2.3 <u>Unitary Representations</u>

A linear space is called *Euclidean* if from each two vectors x
and y of it one can define a function, called the *scalar product* of
x and y, and denoted by (x, y), which satisfies:

(1) $(x, x) \geqslant 0$, $(x, x) = 0$ if and only if $x = 0$;

(2) $(y, x) = \overline{(x, y)}$;

(3) $(\alpha x, y) = \alpha(x, y)$;

(4) $(x_1 + x_2, y) = (x_1, y) + (x_2, Y)$.

One can show that a scalar product can be introduced in every finite-
dimensional space.

An operator D in a finite-dimensional Euclidean space R is
called *unitary* if it preserves the scalar product, namely,
$(Dx, Dy) = (x, y)$ for all x, y of the space R. A representation
$g \to D_g$ is called *unitary* if all its operators D_g are unitary.

In the following six chapters we find the irreducible
representations of the three-dimensional pure rotation group, which
will be denoted by 0_3. This is done by Weyl's method, which makes
use of the homomorphism of the special unitary group of order two, the
group SU_2, onto the rotation group 0_3. The representations are
discussed in terms of two different parameterizations:

(1)   the angle of rotation in a specified direction and the
      spherical angles of the direction of rotation; and

(2)   the traditional Euler angles.

NOTES AND REFERENCES

1. E.P. Wigner, *Group Theory and its Application to the Quantum Mechanics of Atomic Spectra*, Academic Press, New York, 1959.

2. M.A. Naimark, *Linear Representations of the Lorentz Group*, Pergamon, New York, 1964.

3. I.M. Gelfand, M.I. Graev, and N. Ya. Vilenkin, *Generalized Functions, Vol. 5: Integral Geometry and Representation Theory*, Academic Press, New York, 1966.

4. I.M. Gelfand and Z. Ya. Shapiro, *Usp. Mat. Nauk* 7, 3 (1952); *Amer. Math. Soc. Transl. Ser. 2.*, 2, 207 (1956).

5. I.M. Gelfand, R.A. Minlos, and Z. Ya. Shapiro, *Representations of the Rotation and Lorentz Groups and their Applications*, Pergamon Press, Inc., New York, 1963.

6. An operator function D_g is called continuous on a group G if the elements of the matrix of D_g, relative to a fixed basis, are continuous functions on G. This definition of continuity of D_g does not depend on the choice of the basis since the matrix elements relative to another basis are linear combinations, with constant coefficients, of the matrix elements relative to the original basis.

CHAPTER III

THE THREE-DIMENSIONAL PURE ROTATION GROUP

A linear transformation g of the variables x_1, x_2, and x_3, which leaves the form $x_1^2 + x_2^2 + x_3^2$ invariant, is called a *three-dimensional rotation*. The aggregate of all such linear transformations g provides a continuous group, which is isomorphic to the set of all *real orthogonal* 3-dimensional matrices[1] and is known as the *three-dimensional rotation group*. One can easily show that the determinant of every orthogonal matrix is equal to either +1, in which case the transformation describes *pure rotation*, or to -1, in which case it describes a *rotation-reflection*. The aggregate of all pure rotations forms a group, which is a subgroup of the 3-dimensional rotation group, and is known as the *pure rotation group*. *We will be concerned with the 3-dimensional pure rotation group.* This group is denoted by us by 0_3.[2]

§ 3.1 The Euler Angles

Let g be an element of the group 0_3, i.e., a 3-dimensional orthogonal matrix with determinant unity. It is well known that one then can express each such element in terms of a set of three parameters. An example of such parameters is that of the familiar Euler angles, which are defined as the three successive angles of rotation describing the transformation from a given cartesian coordinate system to another one by means of three successive rotations performed in a specific sequence.

The sequence will be started (see Figure 1) by rotating the original system of axes $\vec{x}$ by an angle ϕ_1 clockwise about the z axis.[3] The new coordinate system will be labelled $\vec{\xi}$. One thus has

$$\vec{\xi} = g(\phi_1) \, \vec{x} \, ,$$

where the orthogonal matrix $g(\phi_1)$ is given by

$$g(\phi_1) = \begin{pmatrix} \cos\phi_1 & -\sin\phi_1 & 0 \\ \sin\phi_1 & \cos\phi_1 & 0 \\ 0 & 0 & 1 \end{pmatrix} . \qquad (3.1)$$

In the second stage the intermediate axes $\vec{\xi}$ are rotated about its ξ axis clockwise by an angle θ to another intermediate set which is denoted $\vec{\xi}'$, thus one has

$$\vec{\xi}' = g(\theta) \, \vec{\xi} \, ,$$

where the orthogonal matrix $g(\theta)$ is given by

$$g(\theta) = \begin{pmatrix} 1 & 0 & 0 \\ 0 & \cos\theta & -\sin\theta \\ 0 & \sin\theta & \cos\theta \end{pmatrix} . \qquad (3.2)$$

The ξ' axis is called the *line of nodes*. Finally the $\vec{\xi}'$ axes are rotated clockwise by an angle ϕ_2 about the ξ' axis to produce the desired $\vec{x}'$ system of axes,

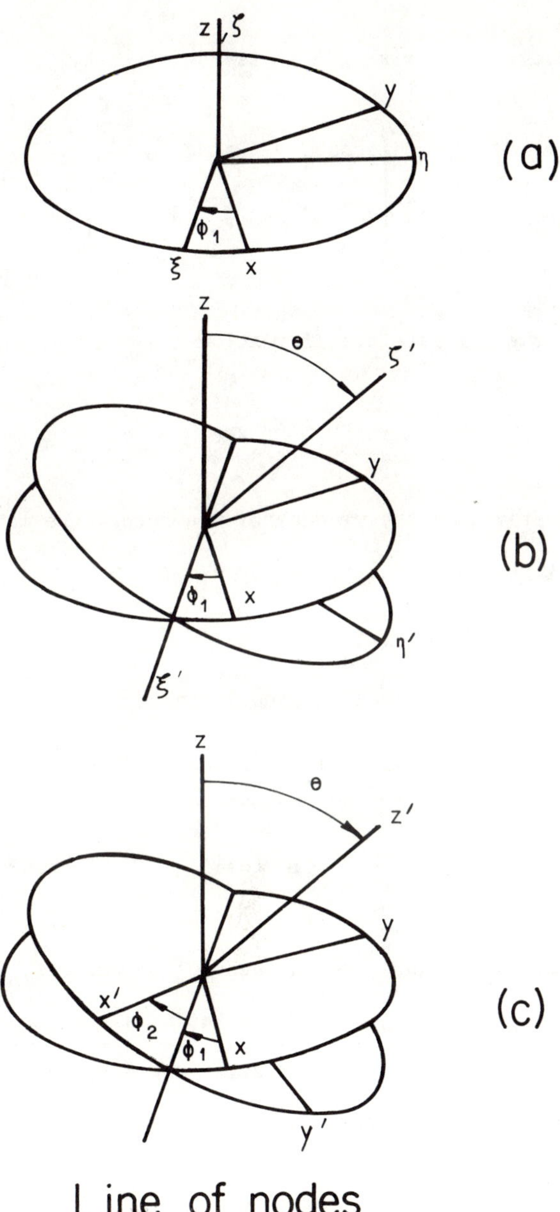

Figure 1. The three rotations defining the Euler angles.

$$\vec{x}' = g(\phi_2) \, \vec{\xi}$$

where the orthogonal matrix $g(\phi_2)$ is now given by

$$g(\phi_2) = \begin{pmatrix} \cos\phi_2 & -\sin\phi_2 & 0 \\ \sin\phi_2 & \cos\phi_2 & 0 \\ 0 & 0 & 1 \end{pmatrix}. \qquad (3.3)$$

The matrix of the complete transformation

$$\vec{x}' = \vec{g}x$$

is given, therefore, by the product of the successive matrices $g = g(\phi_2)\, g(\theta)\, g(\phi_1)$. Hence it is given by

$$g = \begin{pmatrix} \cos\phi_2\cos\phi_1 & -\cos\phi_2\sin\phi_1 & \sin\phi_2\sin\theta \\ -\cos\theta\sin\phi_1\sin\phi_2 & -\cos\theta\cos\phi_1\sin\phi_2 & \\ \sin\phi_2\cos\phi_1 & -\sin\phi_2\sin\phi_1 & -\cos\phi_2\sin\theta \\ +\cos\theta\sin\phi_1\cos\phi_2 & +\cos\theta\cos\phi_1\cos\phi_2 & \\ \sin\theta\sin\phi_1 & \sin\theta\cos\phi_1 & \cos\theta \end{pmatrix}. \qquad (3.4)$$

The angles ϕ_1, θ, ϕ_2 are independent parameters, fully determining the rotation g. They are known as the *Euler Angles*. By their definition, one has $0 \leq \phi_1 \leq 2\pi$, $0 \leq \theta \leq \pi$, and $0 \leq \phi_2 \leq 2\pi$ for the intervals of the angles ϕ_1, θ, and ϕ_2.

NOTES AND REFERENCES

1. A matrix g is called orthogonal if $g^t g = 1$, where g^t is the

transposed of the matrix g.

2. For more details see E.P. Wigner, *Group Theory and its Application to the Quantum Mechanics of Atomic Spectra*, Academic Press, New York, 1959.

3. We use the notation according to which $\vec{x} = (x, y, z) = (x_1, x_2, x_3)$, $\vec{\xi} = (\xi, \eta, \zeta)$, $\vec{\xi}' = (\xi', \eta', \zeta')$, and $\vec{x}' = (x', y', z') = (x_1', x_2', x_3')$.

CHAPTER IV

THE SPECIAL UNITARY GROUP SU_2

Rotations can also be specified by unitary matrices of order two and determinant unity. The aggregate of all such matrices provides a group which is usually denoted by SU_2. The relation between the groups O_3 and SU_2 can be established as follows.[1-4]

§ 4.1 Homomorphism between the Groups O_3 and SU_2

Let x_ℓ and x'_k, with $k, \ell = 1, 2, 3$, denote the coordinates of two cartesian frames related by the transformation

$$x'_k = g_{k\ell} x_\ell \, ,$$

where $g_{k\ell}$ are elements of the matrix $g \, \epsilon \, O_3$. With each coordinate system x_k one associates a 2×2 Hermitian matrix P defined by

$$P = x_k \, \sigma^k \, ,$$

where σ^k are the familiar Pauli spin matrices,

$$\sigma^1 = \begin{pmatrix} 0 & 1 \\ 1 & 0 \end{pmatrix} \, , \quad \sigma^2 = \begin{pmatrix} 0 & i \\ -i & 0 \end{pmatrix} \, , \quad \sigma^3 = \begin{pmatrix} 1 & 0 \\ 0 & -1 \end{pmatrix} \, . \quad (4.1)$$

In terms of the matrix P one requires that the coordinates transform

17

according to the formula

$$P' = uPu^\dagger \ ,$$

where u is an element of the group SU_2,

$$P' = x'_\ell \ \sigma^\ell \ ,$$

and $u^\dagger$ is the Hermitian conjugate of the matrix u. The relations between u of SU_2 and g of O_3 are given by

$$g_{rs} = \frac{1}{2} \, \text{Tr}(\sigma^r u \sigma^s u^\dagger) \ , \tag{4.2}$$

$$u = \mp \, (1 + \sigma^r \sigma^s \, g_{rs})/2(1 + \text{Tr} g)^{\frac{1}{2}} \ , \tag{4.3}$$

where Tr stands for trace.

Accordingly, to each rotation g of the group O_3 there corresponds, by Eq. (4.3), two matrices $\mp u$ of the group SU_2 and, conversely, to each unitary matrix u of the group SU_2 there corresponds, by Eq. (4.2), some rotation g of the group O_3. It thus follows that the group O_3 is homomorphic (see Chapter I) to the group SU_2. For example, the unitary matrices corresponding to the rotations $g(\phi_1)$, $g(\theta)$, and $g(\phi_2)$ given by Equations (3.1)-(3.3) are easily found, using Equation (4.3). They are given by

$$u(\phi_1) = \mp \begin{pmatrix} e^{i\phi_1/2} & 0 \\ 0 & e^{-i\phi_1/2} \end{pmatrix} \ , \tag{4.4}$$

$$u(\theta) = \mp \begin{pmatrix} \cos\frac{\theta}{2} & i\sin\frac{\theta}{2} \\ i\sin\frac{\theta}{2} & \cos\frac{\theta}{2} \end{pmatrix} , \qquad (4.5)$$

and

$$u(\phi_2) = \mp \begin{pmatrix} e^{i\phi_2/2} & 0 \\ 0 & e^{-i\phi_2/2} \end{pmatrix} . \qquad (4.6)$$

A general rotation g, described by the matrix (3.4) will then correspond to the unitary matrix $u = u(\phi_2)\, u(\theta)\, u(\phi_1)$ and is thus given by

$$u = \mp \begin{pmatrix} \cos\frac{\theta}{2}\, e^{i(\phi_2+\phi_1)/2} & i\sin\frac{\theta}{2}\, e^{i(\phi_2-\phi_1)/2} \\ i\sin\frac{\theta}{2}\, e^{-i(\phi_2-\phi_1)/2} & \cos\frac{\theta}{2}\, e^{-i(\phi_2+\phi_1)/2} \end{pmatrix} . \qquad (4.7)$$

NOTES AND REFERENCES

1. M. Carmeli, Representations of the Three-Dimensional Rotation Group in Terms of Direction and Angle of Rotation, *J. Math. Phys.* <u>9</u>, 1987 (1968).

2. E.P. Wigner, *Group Theory and its Application to the Quantum Mechanics of Atomic Spectra*, Academic Press, New York, 1959.

3. M.A. Naimark, *Linear Representations of the Lorentz Group*, Pergamon Press, New York, 1964.

4. H. Goldstein, *Classical Mechanics*, Addison-Wesley Publishing Company, Inc., Reading, Mass., 1965.

CHAPTER V

INVARIANT INTEGRALS OVER THE GROUPS 0_3 and SU_2

A function $y = f(g)$ is said to be defined over the group G if to each element g of G there corresponds a number y.[1-4] If the group is taken to be the rotation group 0_3 and one uses the Euler angles as parameters then $f(g)$, where $g \in 0_3$, becomes simply a function of the angles ϕ_1, θ, ϕ_2, i.e.,

$$f(g) = f(\phi_1, \theta, \phi_2) \ .$$

The function f then satisfies

$$f(\phi_1 + 2\pi, \theta, \phi_2) = f(\phi_1, \theta, \phi_2) \ ,$$

$$f(\phi_1, \theta, \phi_2 + 2\pi) = f(\phi_1, \theta, \phi_2) \ . \tag{5.1}$$

§ 5.1   Invariant Integral over the Group 0_3

The integral $\int f(g)\,dg$ is then called *invariant integral* of the function $f(g)$ over the group 0_3 if it satisfies

$$\int f(g\,g_0)\,dg = \int f(g_0\,g)\,dg$$

$$= \int f(g)\,dg \tag{5.2}$$

21

for any $g_0 \in 0_3$, and

$$\int f(g^{-1})dg = \int f(g)dg \ . \tag{5.3}$$

The expression dg is called a *measure*. When the Euler angles are used to paramaterize the elements g of the group 0_3, one can write dg in terms of the angles ϕ_1, θ, ϕ_2 as

$$dg = (1/8\pi^2) \ \sin\theta \ d\phi_1 \ d\theta \ d\phi_2 \ .$$

One can then easily verify that it satisfies

$$\int dg = 1 \ . \tag{5.4}$$

The integration limits extend over the whole domain of definitions of the variables, i.e., $0 \le \phi_1 \le 2\pi$, $0 \le \theta \le \pi$, and $0 \le \phi_2 \le 2\pi$.

§ 5.2   Invariant Integral over the Group SU_2

The concepts of functions defined over the group 0_3 and invariant integrals defined over the rotation group 0_3 can easily be extended to the special unitary group SU_2. Again, a function $f(u)$ defined over the group SU_2 can be considered as a function of the angles ϕ_1, θ, ϕ_2, i.e.,

$$f(u) = f(\phi_1, \ \theta, \ \phi_2)$$

if the Euler angles are used for parametrization. The analogous periodicity conditions to those of Equation (5.1) for functions defined over 0_3 will now be

$$f(\phi_1 + 4\pi, \theta, \phi_2) = f(\phi_1, \theta, \phi_2) \ ,$$

$$f(\phi_1, \theta, \phi_2 + 4\pi) = f(\phi_1, \theta, \phi_2) \ , \qquad (5.5)$$

$$f(\phi_1 + 2\pi, \theta, \phi_2 + 2\pi) = f(\phi_1, \theta, \phi_2) \ .$$

The invariant integral over the group SU_2 then satisfies

$$\int f(uu_0)\,du = \int f(u_0 u)\,du$$

$$= \int f(u)\,du \qquad (5.6)$$

for any $u \in SU_2$, and

$$\int f(u^{-1})\,du = \int f(u)\,du \ . \qquad (5.7)$$

The measure du can then be expressed in terms of the Euler angles as

$$du = (1/16\pi^2)\ \sin\theta\ d\phi_1\ d\theta\ d\phi_2 \ .$$

It can be shown that it satisfies

$$\int du = 1 \ . \qquad (5.8)$$

The integration limits here will be:

$$0 \leqslant \phi_1 \leqslant 4\pi \ , \quad 0 \leqslant \theta \leqslant \pi \ , \ \text{and} \ \ 0 \leqslant \phi_2 \leqslant 2\pi \ .$$

NOTES AND REFERENCES

1. L. Pontrjagin, *Topological Groups*, Princeton University Press,
Princeton, New Jersey, 1946.

2. A. Weil, *Actualites Sci. Ind.*, No. 869 (1938); *L'integration dans
les groups topologiques et ces applications*, Hermann & Cie.,
Paris, 1940.

3. M.A. Naimark, *Linear Representations of the Lorentz Group*,
Pergamon Press, Inc., New York, 1964.

4. E.P. Wigner, *Group Theory and its Application to the Quantum
Mechanics of Atomic Spectra*, Academic Press, New York, 1959,

CHAPTER VI

REPRESENTATIONS OF THE GROUPS 0_3 AND SU_2

We have seen that the pure rotation group 0_3 is homomorphic to the unimodular unitary group of order two, SU_2, such that to every rotation g of 0_3 there corresponds two matrices $+u$ and $-u$ of SU_2 and, conversely, to every element u of SU_2 there corresponds some rotation g of 0_3.

§ 6.1 Weyl's Method

It thus follows that the description of the representations (see Chapter II) of the group 0_3 is equivalent to that of the group SU_2; a representation $g \rightarrow D_g$ of the group 0_3 is single- or double-valued according to whether or not D_u is equal to D_{-u}. The use of the group SU_2 for finding the representations of the group 0_3 was originally suggested by H. Weyl and has been widely adopted when the Euler angles are used to parameterize the groups. The advantage of Weyl's method is in giving the *double valued* representations along with the proper representations. The double valued representations are important in physical problems dealing with spin-like properties of particles whose spin are half integers.

We point out that, by using Weyl's method, one can obtain a general invariant result that is a function of the element $u \in SU_2$, valid for any parametrization one uses to describe the rotation. To find the

25

representations of the group O_3 in terms of a specific set of parameters, one has merely to express u in terms of these parameters, as is the case when the Euler angles are adopted. In addition, by having the results as functions over the group SU_2, certain relations will be obtained which are invariant under change of the parameters. As an example, the orthogonality relations between the matrix elements of the irreducible representation can be written in the form of an invariant integral over the group SU_2. Hence the relations are valid for any parametrization.

§ 6.2 Infinitesimal Generators

An orthogonal matrix describing a rotation with angle ψ about some direction

$$\vec{n} = (\sin\theta\,\cos\phi,\ \sin\theta\,\sin\phi,\ \cos\theta)$$

is given by[1]

$$g_{rs} = \delta_{rs}\,\cos\psi + n_r n_s (1 - \cos\psi) - \varepsilon_{rst} n^t \sin\psi\ , \qquad (6.1)$$

where r, s, and t take the values from 1 to 3. Rotations $g_1(\psi)$, $g_2(\psi)$, and $g_3(\psi)$ around Ox_1, Ox_2, and Ox_3 axes are then obtained from (6.1) by putting the proper values for the polar angles θ and ϕ.[2] The *infinitesimal matrices*, g_r corresponding to rotations about the axis Ox_r are defined by[3]

$$g_r = \left.\frac{dg_r(\psi)}{d\psi}\right|_{\psi=0}\ , \qquad (6.2)$$

and satisfy the *commutation relations*

$$[g_r, g_s] = \varepsilon_{rst}\, g_t \, , \tag{6.3}$$

where $[g_r, g_s] = g_r g_s - g_s g_r$.

Let us denote a representation of the group 0_3 in an n-dimensional Euclidean space R by $g \to D_g$ and for convenience we denote[4]

$$A_r(\psi) = D_{g_r}(\psi) \, . \tag{6.4}$$

The *basic infinitesimal operators* of the representation are then obtained by

$$A_r = \left. \frac{dA_r(\psi)}{d\psi} \right|_{\psi=0} . \tag{6.5}$$

A representation of the group 0_3 is *uniquely determined* by its basic infinitesimal operators A_r. The determination of all the finite-dimensional representations of the group 0_3 is based on the fact that the operators A_r satisfy the *same* commutation relations that exists among the infinitesimal matrices g_r:

$$[A_r, A_s] = \varepsilon_{rst} A_t \, . \tag{6.6}$$

The operators A_r are skew-Hermitian operators,[5] $A_r^{\dagger} = -A_r$, since, without loss of generality, every finite-dimensional representation of 0_3 can be considered to be unitary.

§ 6.3 <u>Canonical Basis</u>

Defining the new operators

$$L_{\mp} = i\, A_1 \pm A_2 \; ,$$

$$L_3 = i\, A_3 \; ,$$

(6.7)

one then finds for the commutation relations of the infinitesimal

generators L_+, L_-, and L_3 the following:

$$[L_{\mp}, L_3] = \pm L_{\mp} \; ,$$

$$[L_+, L_-] = 2L_3 \; ,$$

$$L_+{}^\dagger = L_- , \; L_3{}^\dagger = L_3 \; .$$

(6.8)

The problem of determining the representation is then reduced to the

determination of the operators $L_{\mp}$ and L_3 satisfying the conditions

(6.8). This is answered by the following: every finite-dimensional

representation of the group O_3 is uniquely determined by a non-

negative integer or half-integer j, the *weight* of the representation.

The space of the representation corresponding to such a number j has

the dimension 2j + 1; the operators $L_{\mp}$ and L_3 of the representation

are given relative to its *canonical basis* f_{-j}, f_{-j+1}, $\ldots$, f_j by

$$L_\pm f_m = [(j \mp m)(j \pm m + 1)]^{\frac{1}{2}} f_{m\pm 1} \; ,$$

$$L_3 f_m = m\, f_m \; ,$$

(6.9)

where $m = -j, -j + 1, \ldots, j.$[6]

§ 6.4   Unitary Matrices Corresponding to Rotations

We now find the unitary matrix u corresponding to the rotation g, of equation (6.1). The matrices u and g are related by Equations (4.2) and (4.3). A direct calculation then gives:

$$u = \bar{+} \begin{pmatrix} \cos\frac{\psi}{2} + i\sin\frac{\psi}{2}\cos\theta & i\sin\frac{\psi}{2}\sin\theta\, e^{i\phi} \\ i\sin\frac{\psi}{2}\sin\theta\, e^{-i\phi} & \cos\frac{\psi}{2} - i\sin\frac{\psi}{2}\cos\theta \end{pmatrix} . \qquad (6.10)$$

This is the unitary matrix $u \in SU_2$ corresponding to a rotation with an angle ψ around the direction $\vec{n}$ specified by the spherical angles θ and ϕ. The corresponding matrix, when the Euler angles are employed, was given in Equation (4.7). It will be noted that

$$u(-\psi,\ \theta,\ \phi) = u^{-1}(\psi,\ \theta,\ \phi) .$$

The unitary matrices $u_1(\psi)$, $u_2(\psi)$, and $u_3(\psi)$ corresponding to the rotations $g_1(\psi)$, $g_2(\psi)$, and $g_3(\psi)$ around the axes of coordinates $0x_1$, $0x_2$, and $0x_3$ (see footnote 2), can be obtained from (6.10) by putting the appropriate values for the angles θ and ϕ. They are:[7]

$$u_1(\psi) = \bar{+} \begin{pmatrix} \cos\frac{\psi}{2} & i\sin\frac{\psi}{2} \\ i\sin\frac{\psi}{2} & \cos\frac{\psi}{2} \end{pmatrix} , \qquad (6.11a)$$

$$u_2(\psi) = \bar{+} \begin{pmatrix} \cos\frac{\psi}{2} & -\sin\frac{\psi}{2} \\ \sin\frac{\psi}{2} & \cos\frac{\psi}{2} \end{pmatrix} , \qquad (6.11b)$$

and

$$u_3(\psi) = \mp \begin{pmatrix} e^{i\psi/2} & 0 \\ 0 & e^{-i\psi/2} \end{pmatrix}. \tag{6.11c}$$

Using these matrices, the operators $A_r(\psi)$ of the group SU_2 will be determined in the next chapter.

NOTES AND REFERENCES

1. The following is essentially based on M. Carmeli, *J. Math. Phys.* $\underline{9}$, 1987 (1968). See also H.E. Moses, *Ann. Phys. (N.Y.)* $\underline{37}$, 224 (1966); $\underline{42}$, 343 (1967); *Nuovo Cimento* $\underline{40A}$, 1120 (1965).

2. These matrices are given by

$$g_1(\psi) = \begin{pmatrix} 1 & 0 & 0 \\ 0 & \cos\psi & -\sin\psi \\ 0 & \sin\psi & \cos\psi \end{pmatrix},$$

$$g_2(\psi) = \begin{pmatrix} \cos\psi & 0 & \sin\psi \\ 0 & 1 & 0 \\ -\sin\psi & 0 & \cos\psi \end{pmatrix},$$

$$g_3(\psi) = \begin{pmatrix} \cos\psi & -\sin\psi & 0 \\ \sin\psi & \cos\psi & 0 \\ 0 & 0 & 1 \end{pmatrix}.$$

3. The g_r are related to $g_r(\psi)$ by $g_r(\psi) = \exp(\psi\, g_r)$, and are given

by

$$g_1 = \begin{pmatrix} 0 & 0 & 0 \\ 0 & 0 & -1 \\ 0 & 1 & 0 \end{pmatrix}, \quad g_2 = \begin{pmatrix} 0 & 0 & 1 \\ 0 & 0 & 0 \\ -1 & 0 & 0 \end{pmatrix}, \quad g_3 = \begin{pmatrix} 0 & -1 & 0 \\ 1 & 0 & 0 \\ 0 & 0 & 0 \end{pmatrix}.$$

4. $A_r(\psi)$ are called the *basic one-parameter groups* of the given representation and define one-parameter groups of operators that satisfy $A_r(\psi_1) A_r(\psi_2) = A_r(\psi_1 + \psi_2)$; they are differentiable functions of ψ and may be expanded as $A_r(\psi) = \exp(\psi A_r)$, where A_r is defined by Equation (6.5).

5. An operator B in a finite-dimensional Euclidean space R is called *adjoint* to the operator A in the same space if $(Ax, y) = (x, By)$ for all x, y of R. The adjoint of an operator A is usually denoted by $A^\dagger$. It can be shown that for any linear operator A there exists one and only one adjoint operator $A^\dagger$, and that the adjoint operator to $A^\dagger$ is A. An operator A is called *Hermitian* if $A^\dagger = A$. An operator A is called *unitary* if and only if $A^\dagger A = 1$.

6. It also follows that for each j there corresponds an irreducible representation of 0_3. If the operators $L_{\mp}$ and L_3 of a representation of 0_3 in a $(2j + 1)$-dimensional space are given relative to some basis $f_{-j}, f_{-j+1}, \ldots, f_j$, then by Equations (6.9) that representation is irreducible.

7. The infinitesimal matrices u_r corresponding to rotations around $0x_r$,

$$u_r = [du_r(\psi)/d\psi]_{\psi=0}$$

are related to the Pauli matrices, Equation (4.1), by

$$u_r = \pm (i/2)\sigma^r.$$

CHAPTER VII

MATRIX ELEMENTS OF IRREDUCIBLE REPRESENTATIONS

A matrix u of the group SU_2 can be considered as that of a linear transformation of the space of all pairs of complex numbers (ξ^1, ξ^2):

$$\xi'^p = \sum_{q=1}^{2} u_{pq} \xi^q \quad (p = 1, 2) . \qquad (7.1)$$

A representation of the group SU_2 can be obtained if one considers several pairs

$$(\xi_1^{\ 1}, \xi_1^{\ 2}), \ \ldots, \ (\xi_k^{\ 1}, \xi_k^{\ 2})$$

and forms all products $\xi_1^{\ p_1} \ldots \xi_1^{\ p_k}$, letting $p_1, \ldots, p_k$ take the values 1, 2, independently. Under the transformation (7.1), this product transforms like

$$\xi'_1^{\ p_1} \ldots \xi'_k^{\ p_k} = \sum_{q_1,\ldots,q_k=1}^{2} u_{p_1 q_1} \ldots u_{p_k q_k} \xi_1^{\ q_1} \ldots \xi_k^{\ q_k} . \qquad (7.2)$$

The product $\xi_1^{\ p_1} \ldots \xi_k^{\ p_k}$ may be considered as a vector in the linear space R_k of all 2^k complex numbers $\xi^{p_1 \ldots p_k}$. The linear transformation $D_u^{(k)}$ of the space R_k is then given by

$$\xi'^{p_1 \cdots p_k} = \sum_{q_1 \cdots, q_k = 1}^{2} u_{p_1 q_1} \cdots u_{p_k q_k} \xi^{q_1 \cdots q_k} . \qquad (7.3)$$

§ 7.1 <u>Spinor Representation</u>

The correspondence $u \rightarrow D_u^{(k)}$ is a representation of the group SU_2, *not* irreducible in general, since the subspace S_k of R_k of all symmetrical vectors ξ is invariant with respect to all the operators $D_u^{(k)}$. The correspondence $u \rightarrow D_u^{(k)}$ is irreducible, however, in the subspace S_k. We denote this representation by Z_k. It is called the *spinor representation* of the group SU_2 and is of weight $k/2$.

An equivalent realization of the representation Z_k is obtained if one identifies the subspace S_k with the $(k + 1)$-dimensional space of homogoneous polynomials $p(z_1, z_2)$ of degree k in the two complex variables z_1 and z_2 and sets up a one-to-one correspondence between ξ of S_k and $p(z_1, z_2)$ in the form

$$p(z_1, z_2) = \sum_{p_1, \cdots, p_k = 1}^{2} \xi^{p_1 \cdots p_k} z_{p_1} \cdots z_{p_k} . \qquad (7.4)$$

The operator $D_u^{(k)}$ for this new realization in the space of polynomials S_k is then given by

$$D_u^{(k)} p(z_1, z_2) = p(z'_1, z'_2) , \qquad (7.5a)$$

where

$$z'_q = \sum_{p=1}^{2} u_{pq} z_p \quad (q = 1, 2) . \qquad (7.5b)$$

Introducing now a new variable $z = z_1/z_2$, the polynomial $p(z_1, z_2)$ can then be written as $z_2^{k} p(z)$, where $p(z)$ is a polynomial in the variable z of degree not exceeding k. The operators $D_u^{(k)}$ of the representation Z_k are, accordingly, given by

$$D_u^{(k)} \, p(z) = (u_{12}z + u_{22})^k \, p\left(\frac{u_{11}z + u_{21}}{u_{12}z + u_{22}}\right). \qquad (7.6)$$

This equation gives, in particular, the operators $A_r(\psi) = D_{u_r(\psi)}$ when the matrices $u_r(\psi)$, Equations (6.11), are used.[1]

§ 7.2 Matrix Elements of Representations

It follows that every irreducible finite-dimensional representation of the group SU_2 is uniquely determined by some non-negative integer or half-integer $j = k/2$, the *weight* of the representation.[2] The functions

$$f_m(z) = \frac{(-z)^{j-m}}{[(j - m)!\,(j + m)!]^{\frac{1}{2}}}, \qquad (7.7)$$

where $m = j, -j + 1, \ldots, j$, then form a canonical basis for the representation Z_k in the space S_k. Using Equation (7.6), one finds

$$D_u^{(k)} \, f_n(z) = \sum_{m=-j}^{j} D^{j}_{mn}(u) \, f_m(z), \qquad (7.8)$$

where $D^{j}_{mn}(u)$ are the matrix elements of the operator D_u of the irreducible representation of weight j relative to the canonical basis, which corresponds to an arbitrary rotation g. Its explicit expression is[3]

$$D^j_{mn}(u) = (-1)^{2j-m-n} \left[\frac{(j-m)!(j+m)!}{(j-n)!(j+n)!}\right]^{\frac{1}{2}}$$

$$\times \Sigma \; \binom{j-n}{a}\binom{j+n}{j-m-a} \; u_{11}^a \; u_{12}^{j-m-a} \; u_{21}^{j-n-a} \; u_{22}^{m+n+a} \; , \qquad (7.9)$$

where the summation runs from $a = \max(0, -m-n)$ to $\min(j-m, j-n)$, and

$$\binom{m}{n} = \frac{m!}{(m-n)!n!} \; .$$

In Equation (7.9) the indices m and n take the values $-j, -j+1, \ldots, j$ and $j = 0, 1/2, 1, 3/2, 2, \ldots$.

To find the matrix elements (7.9) in terms of the variables ψ, θ, and ϕ we simply substitute for u_{pq} their expressions as functions of these variables as given by Equation (6.10).[4] One obtains

$$D^j_{mn}(\psi, \theta, \phi) = (-1)^{2j-m-n} \left[\frac{(j-m)!(j+m)!}{(j-n)!(j+n)!}\right]^{\frac{1}{2}} (i \sin \tfrac{\psi}{2} \sin\theta \; e^{-i\phi})^{m-n}$$

$$\times (\cos \tfrac{\psi}{2} - i \sin \tfrac{\psi}{2} \cos\theta)^{m+n} \; S(j, m, n; x) \; . \qquad (7.10)$$

Here we have used the notation[5]

$$S(j, m, n; x) = 2^{m-j}(j-n)!(j+n)!$$

$$\times \Sigma \; \frac{(x+1)^a \, (x-1)^{j-m-a}}{a!(j-n-a)!(j-m-a)!(a+m+n)!} \; , \qquad (7.11)$$

where x is defined by

$$x = 1 - 2 \sin^2(\psi/2)\sin^2\theta \ .$$

§ 7.3 Properties of $D^j_{mn}(u)$

Finally we discuss the properties of the matrices $D^j_{mn}(u)$.

One first notices that the matrices $D^j(u)$ are unitary. The correspondence $u \to D^j(u)$ is a representation of the group SU_2. Therefore one has

$$D^j_{mn}(u_1\,u_2) = \sum_{n'=-j}^{j} D^j_{mn'}(u_1)\,D^j_{n'n}(u_2) \ .$$

Furthermore, one has

$$D^j(u^{-1}) = [D^j(u)]^{-1} = [D^j(u)]^\dagger \ ,$$

or

$$D^j_{mn}(u) = \bar{D}^j_{nm}(u) \ .$$

Denoting now by γ the unitary matrix

$$\gamma = \begin{pmatrix} e^{-i\psi/2} & 0 \\ 0 & e^{i\psi/2} \end{pmatrix} ,$$

where ψ is a real number. If one applies now the representation formula (7.6), where $p(z)$ is taken as the basis functions $f_m(z)$ of Equation (7.7), one obtains

$$D_\gamma f_m(z) = (-1)^{j-m} e^{ij\psi} \frac{(e^{-i\psi}z)^{j-m}}{\sqrt{(j-m)!(j+m)!}}$$

$$= e^{im\psi} f_m(z) \ .$$

Hence the matrix $D^j(\gamma)$ is diagonal, and $D^j_{nn}(\gamma) = e^{in\psi}$. Furthermore, one easily finds that

$$D^j_{mn}(\gamma u) = e^{im\psi} D^j_{mn}(u) \ ,$$

$$D^j_{mn}(u\gamma) = e^{in\psi} D^j_{mn}(u) \ .$$

We conclude this chapter by giving the orthogonality relation that the matrices D^j satisfy:[6]

$$\int D^{j_1}_{m_1 n_1}(u) \ \bar{D}^{j_2}_{m_2 n_2}(u) \ du = (2j_1 + 1)^{-1} \delta_{j_1 j_2} \delta_{m_1 m_2} \delta_{n_1 n_2} \ . \quad (7.12)$$

NOTES AND REFERENCES

1. For the determination of the operators $A_r(\psi)$, one needs $u_r(\psi)$ only for small values of ψ. The signs in Equations (6.11) are determined by the conditions $\lim u_r(\psi) = 1$ when $\psi \to 0$: hence the $+$ sign must be used.

2. Conversely, for any non-negative integer or half-integer j, there exists an irreducible representation of the group SU_2 of weight j. A representation of weight j can be realized as the spinor representation Z_k, where $k = 2j$; and every finite-dimensional

irreducible representation of the group SU_2 is equivalent to one of the representations Z_k.

3. It will be noted that $D^j_{mn}(-u) = (-1)^{2j} D^j_{mn}(u)$. Thus the representation is single-valued for integer j and double-valued for half-integer j. In the sequel the matrix u of Equation (6.10) will be taken with the $+$ sign.

4. One can easily find the expression of D^j in terms of Euler's angles.

5. It will be noted that the function $S(j, m, n; x)$ is equal to the Jacobi polynomial $p_s^{\alpha\beta}(x)$ when $s = j - \frac{1}{2}(|m + n| + |m - n|)$, $\alpha = |m - n|$, and $\beta = |m + n|$.

6. Relations similar to (7.12) are valid for any compact group. See, for example, L.S. Pontrjagin, *Topological Groups*, Princeton University Press, Princeton, New Jersey, 1946.

CHAPTER VIII

DIFFERENTIAL OPERATORS OF INFINITESIMAL ROTATIONS

We are now in a position to find the differential operators
corresponding to infinitesimal rotations about the coordinate axis,
namely, the operators A_1, A_2, and A_3 and, consequently, the operators
$L_{\mp}$ and L_3. These operators are well known in the literature when the
Euler angles are employed. We here derive these operators in terms of
the variables ψ, θ, and ϕ.

§ 8.1 <u>Representation of O_3 in Space of Functions</u>

Let $g \to D_g$ be an irreducible representation of weight j of the
group O_3 and let $D_{mn} = D^j{}_{mn}$ be its matrix elements. We consider
these elements as functions of the rotation g, $D_{mn} = D_{mn}(g)$. Since
$g \to D_g$ is a representation, one has

$$D_{gg'} = D_g D_{g'} \; .$$

In terms of matrix elements, the last relation is

$$D_{mn}(gg') = \sum_{q=-j}^{j} D_{mq}(g) D_{qn}(g') \; , \qquad (8.1)$$

where $D_{mn}(gg')$ are the matrix elements of the operators $D_{gg'}$. Define
now a transformation U such that

$$U_{g'} \, D_{mn}(g) = D_{mn}(gg') \ . \tag{8.2}$$

Comparing Equations (8.1) and (8.2) we obtain

$$U_{g'} \, D_{mn}(g) = \sum_{q=-j}^{j} D_{qn}(g') \, D_{mq}(g) \ . \tag{8.3}$$

Furthermore, one can show that

$$U_{g'} \, U_{g''} = U_{g'g''} \ . \tag{8.4}$$

It thus follows that the transformation $U_{g'}$ realizes a representation of the group O_3 in the space of $2j + 1$ functions of the mth row of the matrix D_g [compare Equation (7.8)], and that the matrix elements of $U_{g'}$ are $D_{qn}(g')$.[1]

To find the operators A_r we take g' as the rotation through some angle α around the axis $0x_r$ and expand the relation (8.2) in powers of α. Expansion of $D_{mn}(gg')$, which we denote by $D_{mn}(\tilde{\psi}, \tilde{\theta}, \tilde{\phi})$, gives

$$D_{mn}(\tilde{\psi}, \tilde{\theta}, \tilde{\phi}) = D_{mn}(\psi, \theta, \phi)$$

$$+ \alpha \left[\frac{\partial D_{mn}}{\partial \psi} \frac{d\tilde{\psi}}{d\alpha} + \frac{\partial D_{mn}}{\partial \theta} \frac{d\tilde{\theta}}{d\alpha} + \frac{\partial D_{mn}}{\partial \phi} \frac{d\tilde{\phi}}{d\alpha} \right]_{\alpha=0} + \ \dots \ . \tag{8.5}$$

To determine the infinitesimal operators A_r we have to determine the functions

$$\left. \frac{d\tilde{\psi}}{d\alpha} \right|_{\alpha=0}, \quad \left. \frac{d\tilde{\theta}}{d\alpha} \right|_{\alpha=0}, \quad \text{and} \quad \left. \frac{d\tilde{\phi}}{d\alpha} \right|_{\alpha=0}, \tag{8.6}$$

for each rotation.

Now the matrix of the rotation g is a function of the angles ψ, θ, and ϕ which, by Equation (6.1), has the form

$$
\begin{pmatrix}
\cos\psi & \sin^2\theta\cos\phi\sin\phi(1-\cos\psi) & \sin\theta\cos\theta\cos\phi(1-\cos\psi) \\
+ \sin^2\theta\cos^2\phi(1-\cos\psi) & - \cos\theta\sin\psi & + \sin\theta\sin\phi\sin\psi \\
\\
\sin^2\theta\sin\phi\cos\phi(1-\cos\psi) & \cos\psi & \sin\theta\cos\theta\sin\phi(1-\cos\psi) \\
+ \cos\theta\sin\psi & + \sin^2\theta\sin^2\phi(1-\cos\psi) & - \sin\theta\cos\phi\sin\psi \\
\\
\sin\theta\cos\theta\cos\phi(1-\cos\psi) & \sin\theta\cos\theta\sin\phi(1-\cos\psi) & \cos\psi \\
- \sin\theta\sin\phi\sin\psi & + \sin\theta\cos\phi\sin\psi & + \cos^2\theta(1-\cos\psi)
\end{pmatrix}.
$$

$$(8.7)$$

The matrix of rotation gg' is given by some angles $\tilde{\psi}$, $\tilde{\theta}$, and $\tilde{\phi}$ which depend on the rotation angle α and which are equal to ψ, θ, and ϕ when $\alpha = 0$. It will also be noted that expansion of the matrix gg' in a power series in α gives

$$
gg' = g(\psi,\ \theta,\ \phi) + \alpha\ \{\frac{\partial g}{\partial\psi}\ \frac{d\tilde{\psi}}{d\alpha}\bigg|_{\alpha=0}
$$

$$
+ \frac{\partial g}{\partial\theta}\ \frac{d\tilde{\theta}}{d\alpha}\bigg|_{\alpha=0} + \frac{\partial g}{\partial\phi}\ \frac{d\tilde{\phi}}{d\alpha}\bigg|_{\alpha=0}\} + \ldots . \qquad (8.8)
$$

To find the infinitesimal operator A_1 we identify g' with the rotation with angle α around $0x_1$ given by

$$
g_1(\alpha) = \begin{pmatrix}
1 & 0 & 0 \\
0 & \cos\alpha & -\sin\alpha \\
0 & \sin\alpha & \cos\alpha
\end{pmatrix} .
$$

Therefore

$$g_1(\alpha) = \begin{pmatrix} 1 & 0 & 0 \\ 0 & 1 & 0 \\ 0 & 0 & 1 \end{pmatrix} + \alpha \begin{pmatrix} 0 & 0 & 0 \\ 0 & 0 & -1 \\ 0 & 1 & 0 \end{pmatrix} + \dots . \tag{8.9}$$

Therefore we obtain for the product of g with g_1:

$$gg_1 = g(\psi, \theta, \phi) + \alpha \begin{pmatrix} 0 & g_{13} & -g_{12} \\ 0 & g_{23} & -g_{22} \\ 0 & g_{33} & -g_{32} \end{pmatrix} + \dots . \tag{8.10}$$

On the other hand, gg_1 is given by Equation (8.8) when $g_1 = g'$.
Comparing these two expressions for gg_1, we obtain equations from which
the three expressions given in (8.6) can be determined for the case of
rotation about Ox_1. We obtain[2]

$$2\sin\theta\cos\phi\sin\frac{\psi}{2}\left(-\sin\theta\,\sin\phi\left.\frac{d\tilde{\phi}}{d\alpha}\right|_{\alpha=0} + \cos\theta\,\cos\phi\left.\frac{d\tilde{\theta}}{d\alpha}\right|_{\alpha=0}\right)$$

$$- \cos\frac{\psi}{2}(1 - \sin^2\theta\,\cos^2\phi)\left.\frac{d\tilde{\psi}}{d\alpha}\right|_{\alpha=0} = 0 , \tag{8.11a}$$

$$2\sin\theta\,\sin\phi\,\sin\frac{\psi}{2}\left(\sin\theta\,\cos\phi\left.\frac{d\tilde{\phi}}{d\alpha}\right|_{\alpha=0} + \cos\theta\,\sin\phi\left.\frac{d\tilde{\theta}}{d\alpha}\right|_{\alpha=0}\right)$$

$$- \cos\frac{\psi}{2}(1 - \sin^2\theta\,\sin^2\phi)\left.\frac{d\tilde{\psi}}{d\alpha}\right|_{\alpha=0}$$

$$= \sin\theta\left(\cos\theta\,\sin\phi\,\sin\frac{\psi}{2} - \cos\phi\,\cos\frac{\psi}{2}\right) , \tag{8.11b}$$

$$2\cos\theta \, \sin\frac{\psi}{2} \, \frac{d\tilde{\theta}}{d\alpha}\bigg|_{\alpha=0} + \cos\frac{\psi}{2} \, \sin\theta \, \frac{d\tilde{\psi}}{d\alpha}\bigg|_{\alpha=0}$$

$$= \cos\theta \, \sin\phi \, \sin\frac{\psi}{2} + \cos\phi \, \cos\frac{\psi}{2} \,. \tag{8.11c}$$

§ 8.2 Angular Momentum Operators

The solution of Equations (8.11) can easily be shown to be[3]

$$\frac{d\tilde{\psi}}{d\alpha}\bigg|_{\alpha=0} = \cos\phi \, \sin\theta \,,$$

$$\frac{d\tilde{\theta}}{d\alpha}\bigg|_{\alpha=0} = \frac{1}{2}\left(\sin\phi + \cos\frac{\psi}{2}\cos\theta\cos\phi\right) \,, \tag{8.12}$$

$$\frac{d\tilde{\phi}}{d\alpha}\bigg|_{\alpha=0} = \frac{1}{2}\cosec\theta \left(\cos\theta\cos\phi - \cot\frac{\psi}{2}\sin\phi\right) \,.$$

Using Equations (8.12) in Equation (8.5), we find the operator A_1 corresponding to the rotation around Ox_1:

$$A_1 = \cos\phi \, \sin\theta \, \frac{\partial}{\partial\psi} + \frac{1}{2}\left(\sin\phi + \cot\frac{\psi}{2}\cos\theta\cos\phi\right)\frac{\partial}{\partial\theta}$$

$$+ \frac{1}{2}\cosec\theta\left(\cos\theta\cos\phi - \cot\frac{\psi}{2}\sin\phi\right)\frac{\partial}{\partial\phi} \,. \tag{8.13a}$$

The operators A_2 and A_3 are found in a similar way:

$$A_2 = \sin\phi \, \sin\theta \, \frac{\partial}{\partial\psi} - \frac{1}{2}\left(\cos\phi - \cot\frac{\psi}{2}\cos\theta\sin\phi\right)\frac{\partial}{\partial\theta}$$

$$+ \frac{1}{2}\cosec\theta\left(\cos\theta\sin\phi + \cot\frac{\psi}{2}\cos\phi\right)\frac{\partial}{\partial\phi} \,, \tag{8.13b}$$

$$A_3 = \cos\theta \, \frac{\partial}{\partial\psi} - \frac{1}{2} \cot \frac{\psi}{2} \sin\theta \, \frac{\partial}{\partial\theta} - \frac{\partial}{\partial\phi} \, . \qquad (8.13c)$$

Using the last three equations in Equation (6.7) one obtains for the operators L_+, L_-, and L_3:

$$L_\pm = i \, e^{\pm i\phi} \, \{ \sin\theta \, \frac{\partial}{\partial\psi} + \frac{1}{2} \, (\mp i + \cot \frac{\psi}{2} \cos\theta) \, \frac{\partial}{\partial\theta}$$

$$+ \frac{1}{2} \, \mathrm{cosec}\,\theta \, (\cos\theta \pm i \cot \frac{\psi}{2}) \, \frac{\partial}{\partial\phi} \} \, , \qquad (8.14a)$$

$$L_3 = i \, (\cos\theta \, \frac{\partial}{\partial\psi} - \frac{1}{2} \cot \frac{\psi}{2} \sin\theta \, \frac{\partial}{\partial\theta} - \frac{\partial}{\partial\phi}) \, . \qquad (8.14b)$$

The operators derived above were expressed in terms of the angle of rotation ψ and the spherical angles of direction of rotation θ and ϕ. One can use, however, the Euler angles and obtain the standard expressions of angular momentum operators given in other treaties.[4,5]

In the following the Lorentz group is introduced along with its infinitesimal matrices and basic infinitesimal operators. The commutation relations that these matrices and operators satisfy are also given. This is the infinitesimal approach to finding the representation of the Lorentz group. Each representation is shown to be determined by a pair of numbers that completely determine the representation.

NOTES AND REFERENCES

1. The representation $g' \to U_{g'}$ in the space of functions $D_{mq}(g)$, $q = -j, -j + 1, \ldots, j$, is irreducible, and the $D_{mq}(g)$ form a

canonical basis in this space. Hence the operators $L_\mp$ and L_3 of this representation satisfy the relation (1.28), i.e.,

$$L_\pm \, D^j_{mn}(g) = [(j \pm n + 1)(j \mp n)]^{\frac{1}{2}} \, D^j_{m,n\pm1}(g) \; ,$$

$$L_3 \, D^j_{mn}(g) = n \, D^j_{mn}(g) \; .$$

2. One obtains nine equations; only three of them are independent. Equations (8.11) are obtained by equating the diagonal element of the matrices (8.8) and (8.10).

3. M. Carmeli, Representations of the Three-Dimensional Rotation Group in Terms of Direction and Angle of Rotation, *J. Math. Phys.* <u>9</u>, 1987 (1968).

4. E.P. Wigner, *Group Theory and its Application to the Quantum Mechanics of Atomic Spectra*, Academic Press, New York, 1959.

5. M.A. Naimark, *Linear Representations of the Lorentz Group*, Pergamon Press, New York, 1964.

CHAPTER IX

THE LORENTZ GROUP

A linear transformation g of the variables[1] x_1, x_2, x_3 and x_4 which leaves the form

$$x_1^2 + x_2^2 + x_3^2 - x_4^2$$

invariant is called a *Lorentz transformation*. The aggregate of all such linear, homogeneous, transformations g provides a continuous group which is known as the *Lorentz group*.

Let x_μ and x_ν', with μ, ν = 1, 2, 3, 4, denote the coordinates of two Lorentz frames that are related by the transformation $x_\mu' = g_{\mu\nu} x_\nu$. The functions $g_{\mu\nu}$ are elements of the matrix g of the Lorent group. Then the square of the lengths in the two systems are equal:

$$x_1'^2 + x_2'^2 + x_3'^2 - x_4'^2 = x_1^2 + x_2^2 + x_3^2 - x_4^2 \ .$$

As a result of this equation one has

$$\sum_{i=1}^{3} (g_{i\lambda} x_\lambda)^2 - (g_{4\lambda} x_\lambda)^2 = \sum_{i=1}^{3} x_i^2 - x_4^2 \ . \tag{9.1}$$

Equating coefficients of $x_\mu x_\nu$ of both sides of Equation (9.1) then gives the relation

$$\sum_{i=1}^{3} g_{i\mu}g_{i\nu} - g_{4\mu}g_{4\nu} = \eta_{\mu\nu} \, , \tag{9.2}$$

where $\eta_{\mu\nu}$ (and later on $\eta^{\mu\nu}$) is the flat space metric given by the matrix

$$\eta = \begin{pmatrix} 1 & 0 & 0 & 0 \\ 0 & 1 & 0 & 0 \\ 0 & 0 & 1 & 0 \\ 0 & 0 & 0 & -1 \end{pmatrix} . \tag{9.3}$$

Using matrix notation, Equation (9.2) can be written in the form

$$g^{t} \, \eta \, g = \eta \, , \tag{9.4}$$

where g^{t} is the transposed matrix to g. Equation (9.4) shows that

$$(\det \, g)^{2} = 1 \, ,$$

and accordingly the determinant of every Lorentz transformation g is equal to either $+1$, in which case the transformation is called *proper*, or to -1, in which case the transformation is called *improper*. This is similar to the case of the rotation group discussed in previous chapters.

§ 9.1 Orthochronous Lorentz Transformations

From Equation (9.2), when one takes $\mu = \nu = 4$, one obtains

$$g_{14}^{2} + g_{24}^{2} + g_{34}^{2} - g_{44}^{2} = -1 \, . \tag{9.5}$$

Hence one has $g_{44}^{2} \geq 1$, and therefore $g_{44} \geq 1$ or $g_{44} \leq 1$. A Lorentz transformation whose element $g_{44} \geqslant 1$ is called *orthochronous*. The aggregate of all orthochronous Lorentz transformations provides a subgroup of the Lorentz group.[2] The aggregate of all proper, orthochronous Lorentz transformations also provides a group which is a subgroup of the Lorentz group.[3-5]

In the following we will be concerned with the group of all *proper, orthochronous, Lorentz transformations*. This group will be denoted by us by L.

NOTES AND REFERENCES

1. The coordinate $x_4 = ct$, where c is the velocity of light and t is the time.

2. A Lorentz transformation satisfies the condition $g_{44} \geqslant 1$ if and only if it transforms every positive time-like vector into another positive time-like vector. (A vector V_μ is called *time-like* if $V_\mu V^\mu = V_\mu \eta^{\mu\nu} V_\mu = V^2 < 0$. A time-like vector V_μ is called *positive* or *negative* according to whether $V_4 > 0$ or $V_4 < 0$).

3. For details see R.F. Streater and A.S. Wightman, *PCT, Spin and Statistics and All That*, W.A. Benjamin, New York, 1964.

4. W. Rühl, *The Lorentz Group and Harmonic Analysis*, W.A. Benjamin, Inc., New York, 1970.

5. F.R. Halpern, *Special Relativity and Quantum Mechanics*, Prentice-Hall, Inc., Englewood Cliffs, N.J., 1968.

CHAPTER X

THE INFINITESIMAL APPROACH

§ 10.1 Infinitesimal Lorentz Matrices

Rotations $a_1(\psi)$, $a_2(\psi)$, $a_3(\psi)$ and Lorentz transformations (boost) $b_1(\psi)$, $b_2(\psi)$, $b_3(\psi)$, around and along $0x_1$, $0x_2$, $0x_3$ can then be written explicitly.[1] As in the case of the rotation group, the infinitesimal matrices a_r and b_r of the group L are defined by[2]

$$a_r = \left.\frac{da_r(\psi)}{d\psi}\right|_{\psi=0} , \qquad b_r = \left.\frac{db_r(\psi)}{d\psi}\right|_{\psi=0} , \qquad (10.1)$$

and are easily seen to satisfy the commutation relations

$$[a_i, a_j] = \varepsilon_{ijk}a_k ,$$

$$[b_i, b_j] = -\varepsilon_{ijk}a_k , \qquad (10.2)$$

$$[a_i, b_j] = \varepsilon_{ijk}b_k .$$

Here ε_{ijk} is the usual Levi-Civita symbol defined by $\varepsilon_{123} = 1$.

§ 10.2 Infinitesimal Operators

We denote an arbitrary linear representation of the group L in an infinite-dimensional space B by $g \to D_g$ and for convenience we put[3]

$$A_r(\psi) = D_{a_r(\psi)} \quad , \quad B_r(\psi) = D_{b_r(\psi)} \ . \tag{10.3}$$

The basic infinitesimal operators of the one-parameter groups $A_r(\psi)$ and $B_r(\psi)$ are then defined by[4]

$$A_r = \left. \frac{dA_r(\psi)}{d\psi} \right|_{\psi=0} \quad , \quad B_r = \left. \frac{dB_r(\psi)}{d\psi} \right|_{\psi=0} \ , \tag{10.4}$$

if the representation is finite-dimensional. If the representation $g \to D_g$ is infinite-dimensional, however, the operator functions $A_r(\psi)$ and $B_r(\psi)$ might be non-differentiable, but there may still exist a vector x for which $A_r(\psi)x$ and $B_r(\psi)x$ are differentiable vector-functions.[5]

A representation $g \to D_g$ of the group L is completely determined by its infinitesimal operators A_i and B_i, $i = 1, 2, 3$. The determination of the irreducible representations of the group L is based on the fact that the basic infinitesimal operators of a representation satisfy the same commutation relations that exists among the infinitesimal matrices a_r and b_r, namely:

$$[A_i, A_j] = \epsilon_{ijk}A_k \ ,$$

$$[B_i, B_j] = -\epsilon_{ijk}A_k \ , \tag{10.5}$$

$$[A_i, B_j] = \epsilon_{ijk}B_k \ .$$

Defining now the new infinitesimal operators

$$L_{\mp} = iA_1 \pm A_2 \, , \quad L_3 = iA_3 \, ,$$

$$(10.6)$$

$$F_{\mp} = iB_1 \pm B_2 \, , \quad F_3 = iB_3 \, ,$$

one then finds that they satisfy the following commutation relations:

$$[L_{\mp}, L_3] = \pm L_{\mp} \, , \quad [L_+, L_-] = 2L_3 \, ,$$

$$[F_{\mp}, F_3] = \mp L_{\mp} \, , \quad [F_+, F_-] = - 2L_3 \, ,$$

$$[L_{\pm}, F_{\pm}] = 0 \, , \quad [L_3, F_3] = 0 \, , \qquad (10.7)$$

$$[L_{\pm}, F_3] = \mp F_{\pm} \, , \quad [F_{\pm}, L_3] = \mp F_{\pm} \, ,$$

$$[L_{\pm}, F_{\pm}] = \pm 2F_3 \, .$$

The problem of determining a representation then reduces to the determination of $L_{\pm}$, L_3, $F_{\pm}$, F_3 satisfying the conditions (10.7).

Now, since the three-dimensional pure rotation group 0_3 is a subgroup of the proper, orthochronous Lorentz group L, obviously every representation of the group L is also a representation of the group 0_3. Clearly, if a given representation of L is irreducible it needs not be irreducible when considered as a representation of the group 0_3. In fact, any infinite-dimensional representation of the group L, when regarded as a representation of the group 0_3, is highly reducible; it is equivalent to a direct sum of an infinite number of irreducible representations. The space of representation R of any irreducible representation of the group L is, therefore, a closed direct sum of subspaces M^j, where M^j is the $(2j + 1)$-dimensional space in which

the irreducible representation of weight j of the group O_3 is
realized.

Following the standard convention, one chooses the $2j + 1$
normalized eigenvectors of the operator L_3 as the canonical basis for
the subspace M^j. Let these base vectors be denoted as f^j_m, where
$m = -j, -j + 1, \ldots, j$, the superscript j indicates the subspace to
which f^j_m belongs,[6] and the subscript is the eigen-value of the
operator L_3. A detailed investigation of the commutation relations
(10.7) in terms of the canonical basis f^j_m then leads to the following
conclusions:

(a) Each irreducible representation of the group L is
     characterized by a pair of numbers (j_0, c), where j_0 is
     integral or half-integral, and c is a complex number.

(b) The space $R(j_0, c)$ of any given irreducible infinite-
     dimensional representation of the group L is characterized
     by the integer or half-integer j_0 such that

$$R(j_0, c) = M^{j_0} \oplus M^{j_0+1} \oplus \ldots .$$

     The whole space $R(j_0, c)$ is apanned, therefore, by the set
     of base vectors f^j_m, where $j = j_0, j_0 + 1, j_0 + 2, \ldots$, and
     $m = -j, -j + 1, \ldots, j$. If the given irreducible
     representation is finite-dimensional then the direct sum of
     the subspaces M's terminates after a finite number of terms.

(c) A given representation is finite-dimensional if and only if

$$c^2 = (j_0 + n)^2 ,$$

for some natural number n.

(d) The irreducible representation corresponding to a given pair
$(j_0,\ c)$ is, with a suitable choice of basis f_m^j in the space
of representation, given by the formulas[7]

$$L_{\pm}\ f_m^j = [(j \pm m + 1)(j \mp m)]^{\frac{1}{2}}\ f_{m\pm1}^j \ ,$$

$$L_3\ f_m^j = m\ f_m^j \ ,$$

$$F_{\pm}\ f_m^j = \pm\ [(j \mp m)(j \mp m - 1)]^{\frac{1}{2}}\ C_j\ f_{m\pm1}^{j-1}$$

$$- [(j \pm m)(j \pm m + 1)]^{\frac{1}{2}}\ A_j\ f_{m\pm1}^j \qquad (10.8)$$

$$\pm [(j \pm m + 1)(j \pm m + 2)]^{\frac{1}{2}}\ C_{j+1}\ f_{m\pm1}^{j+1} \ ,$$

$$F_3\ f_m^j = [(j - m)(j + m)]^{\frac{1}{2}}\ C_j\ f_m^{j-1}\ -mA_j\ f_m^j$$

$$- [(j + m + 1)(j - m + 1)]^{\frac{1}{2}}\ C_{j+1}\ f_m^{j+1} \ .$$

Here we have used the symbols A_j and C_j which are defined
by

$$A_j = i\ c\ j_0/j(j + 1) \ ,$$

and

$$C_j = i(j^2 - j_0^2)^{\frac{1}{2}}\ (j^2 - c^2)^{\frac{1}{2}}/j(4j^2 - 1)^{\frac{1}{2}} \ .$$

(e) To each pair of numbers $(j_0,\ c)$, where j_0 is integral or
half-integral and c is complex, there corresponds a

representation $g \to D_g$ of the group L, whose infinitesimal operators are given by Equations (10.8).

§ 10.3 Unitarity Conditions

If the representation $g \to D_g$ of the group L is unitary,[8,9] then Equations (10.8) satisfy certain conditions which are summarized below.

Let A be an infinitesimal operator of a unitary representation $g \to D_g$ of the group L. Then $A(t) = D_{a(t)}$ is a unitary operator and therefore its adjoint satisfies:[10]

$$[A(t)]^{\dagger} = A(-t) .$$

Accordingly one has

$$(A(t)f, \, g) = (f, \, A(-t)g) .$$

Differentiating both sides of this equation with respect to t we obtain for $t = 0$,

$$(Af, \, g) = - (f, \, Ag) . \qquad (10.9)$$

Using this relation one then easily finds that

$$(L_+ f, \, g) = (f, \, L_- \, g) ,$$

$$(L_3 f, \, g) = (f, \, L_3 g) ,$$

$$(F_+ f, \, g) = (f, \, F_- \, g) , \qquad (10.10)$$

$$(F_3 f, \, g) = (f, \, F_3 g) .$$

A systematic use of Equation (10.10) in (10.8) then leads to the following.

If the irreducible representation $g \to D_g$ of the group L is unitary then the pair (j_0, c) characterizing it satisfies either: (a) c is purely imaginary and j_0 is an arbitrary non-negative integral or half-integral number; or (b) c is a real number in the intervals $0 < |c| \leq 1$ and $j_0 = 0$.

The representations corresponding to case (a) are called the *principal series of representations* and those corresponding to case (b) are called the *complementary series*.

NOTES AND REFERENCES

1. These matrices are given by

$$a_1(\psi) = \begin{pmatrix} 1 & 0 & 0 & 0 \\ 0 & \cos\psi & -\sin\psi & 0 \\ 0 & \sin\psi & \cos\psi & 0 \\ 0 & 0 & 0 & 1 \end{pmatrix},$$

$$a_2(\psi) = \begin{pmatrix} \cos\psi & 0 & \sin\psi & 0 \\ 0 & 1 & 0 & 0 \\ -\sin\psi & 0 & \cos\psi & 0 \\ 0 & 0 & 0 & 1 \end{pmatrix},$$

$$a_3(\psi) = \begin{pmatrix} \cos\psi & -\sin\psi & 0 & 0 \\ \sin\psi & \cos\psi & 0 & 0 \\ 0 & 0 & 1 & 0 \\ 0 & 0 & 0 & 1 \end{pmatrix} ,$$

and

$$b_1(\psi) = \begin{pmatrix} \cosh\psi & 0 & 0 & \sinh\psi \\ 0 & 1 & 0 & 0 \\ 0 & 0 & 1 & 0 \\ \sinh\psi & 0 & 0 & \cosh\psi \end{pmatrix} ,$$

$$b_2(\psi) = \begin{pmatrix} 1 & 0 & 0 & 0 \\ 0 & \cosh\psi & 0 & \sinh\psi \\ 0 & 0 & 1 & 0 \\ 0 & \sinh\psi & 0 & \cosh\psi \end{pmatrix} ,$$

$$b_3(\psi) = \begin{pmatrix} 1 & 0 & 0 & 0 \\ 0 & 1 & 0 & 0 \\ 0 & 0 & \cosh\psi & \sinh\psi \\ 0 & 0 & \sinh\psi & \cosh\psi \end{pmatrix} .$$

2. The a_r and b_r are related to $a_r(\psi)$ and $b_r(\psi)$ by

$$a_r(\psi) = \exp(\psi a_r) , \qquad b_r(\psi) = \exp(\psi b_r) ,$$

and are given by

$$a_1 = \begin{pmatrix} 0 & 0 & 0 & 0 \\ 0 & 0 & -1 & 0 \\ 0 & 1 & 0 & 0 \\ 0 & 0 & 0 & 0 \end{pmatrix},$$

$$a_2 = \begin{pmatrix} 0 & 0 & 1 & 0 \\ 0 & 0 & 0 & 0 \\ -1 & 0 & 0 & 0 \\ 0 & 0 & 0 & 0 \end{pmatrix},$$

$$a_3 = \begin{pmatrix} 0 & -1 & 0 & 0 \\ 1 & 0 & 0 & 0 \\ 0 & 0 & 0 & 0 \\ 0 & 0 & 0 & 0 \end{pmatrix},$$

and

$$b_1 = \begin{pmatrix} 0 & 0 & 0 & 1 \\ 0 & 0 & 0 & 0 \\ 0 & 0 & 0 & 0 \\ 1 & 0 & 0 & 0 \end{pmatrix},$$

$$b_2 = \begin{pmatrix} 0 & 0 & 0 & 0 \\ 0 & 0 & 0 & 1 \\ 0 & 0 & 0 & 0 \\ 0 & 1 & 0 & 0 \end{pmatrix},$$

$$b_3 = \begin{pmatrix} 0 & 0 & 0 & 0 \\ 0 & 0 & 0 & 0 \\ 0 & 0 & 0 & 1 \\ 0 & 0 & 1 & 0 \end{pmatrix}.$$

3. $A_r(\psi)$ and $B_r(\psi)$ are continuous functions of ψ and are called basic one-parameter groups of operators for the given representation. They satisfy the relations $A_r(\psi_1)\,A_r(\psi_2) = A_r(\psi_1 + \psi_2)$, $B_r(\psi_1)\,B_r(\psi_2) = B_r(\psi_1 + \psi_2)$. $A_r(0) = 1$, $B_r(0) = 1$. If the representation is finite-dimensional then the operators $A_r(\psi)$ and $B_r(\psi)$ are differentiable functions of ψ. If the representation is infinite-dimensional, however, these operators might be non-differentiable.

4. $A_r(\psi)$ and $B_r(\psi)$ might then be expanded in terms of A_r and B_r as $A_r(\psi) = \exp(\psi A_r)$, $B_r(\psi) = \exp(\psi B_r)$.

5. In general, let $A(t)$ be a continuous one-parameter group of operators in a Banach space R, and denote by $X(A)$ the set of all vectors $x \in R$ for which the limit of $(A(t)\,x - x)/t$, when $t \to 0$, exists in the sense of the norm in R. Obviously the set $X(A)$ contains the vector $x = 0$. Define now the operator A for all $x \in X(A)$ by $Ax = \lim\{A(t)\,x - x)/t\}$ at the limit $t \to 0$. The domain of definition, $X(A)$, of the operator A is a subspace of R, and A is linear, i.e., $A(\lambda_1 x_1 + \lambda_2 x_2) = \lambda_1\,A\,x_1 + \lambda_2\,A\,x_2$ for x_1, $x_2 \in X(A)$. Such an operator A is called the infinitesimal operator of the one-parameter group $A(t)$. If $A(t) = D_a(t)$ is the group of operators of the representation $g \to D_g$, corresponding to a one-parameter subgroup $a(t)$ of the group L, the corresponding

operator A is then called the *infinitesimal operator* of the
representation $g \to D_g$. For more details see, for example,
M. Carmeli and S. Malin, Finite- and Infinite-Dimensional
Representations of the Lorentz Group, *Fortschritte der Physik* 21,
397 (1973); M.A. Naimark, *Linear Representations of the Lorentz
Group*, Pergamon Press, New York, 1964.

6. The superscript in f_m^j specifies the subspace uniquely since each
irreducible representation of 0_3 is contained at most once in any
given irreducible representation of the group L.

7. Equations (10.8), for the unitary representations case and under
certain assumptions, were first obtained by Gelfand (see
M.A. Naimark, *Linear Representations of the Lorentz Group*, Pergamon,
New York, 1964, p. 117); they later on were rederived by Harish-
Chandra, *Proc. Roy. Soc.* A189, 372 (1947) and *Phys. Rev.* 71, 793
(1947), and by I.M. Gelfand and A.M. Iaglom, *Zh. Eksp. Theor. Fiz.*
18, 703 (1948).

8. For the physical significance of non-unitary representations see
A.O. Barut and S. Malin, *Revs. Mod. Phys.* 40, 632 (1968); *Nuovo
Cimento* 58A, 835 (1968).

9. A representation $g \to D_g$ of a group G in a space R is called
unitary if R is a Hilbert space and D_g is a unitary operator for
all $g \in G$. This implies that $(D_g x, D_g y) = (x, y)$ for all $g \in G$
and all $x, y \in R$, where (x, y) denotes the scalar product in R.

10. An operator B is called an adjoint to the operator A if
$(Ax, y) = (x, By)$ for all $x, y \in R$.

CHAPTER XI

SPINOR REPRESENTATION OF THE LORENTZ GROUP

§ 11.1 The Group SL(2, C) and the Lorentz Group

In what follows we will use the fact that elements of the proper, orthochronous, Lorentz group L can be described by means of elements of SL(2, C), the group of all 2×2 complex matrices with determinant unity. The relation between these two groups resembles that between the groups O_3 and SU_2 and can be established as follows.

Let x_α and x'_β, with $\alpha, \beta = 1, 2, 3, 4,$ describe the coordinates of two Lorentz frames, thus they are related by the coordinate transformation

$$x'_\alpha = g_{\alpha\beta}\, x_\beta \, , \qquad\qquad (11.1)$$

where $g_{\alpha\beta} \in L$. One associates with each coordinate system x a 2×2 Hermitian matrix Q defined by

$$Q = x_\beta\, \sigma^\beta \, , \qquad\qquad (11.2)$$

where σ^k, k = 1, 2, 3, are the Pauli spin matrices,

$$\sigma^1 = \begin{pmatrix} 0 & 1 \\ 1 & 0 \end{pmatrix} , \quad \sigma^2 = \begin{pmatrix} 0 & i \\ -i & 0 \end{pmatrix} , \quad \sigma^3 = \begin{pmatrix} 1 & 0 \\ 0 & -1 \end{pmatrix} , \qquad (11.3)$$

65

and σ^4 is the 2×2 unit matrix. In terms of the Q's one demands that the coordinate transformation (11.1) be expressed as

$$Q' = a \, Q \, a^\dagger \, , \qquad\qquad (11.4)$$

where a is an element of SL(2, C),

$$Q' = x'_\beta \, \sigma^\beta \, ,$$

and $a^\dagger$ is the Hermitian conjugate of a. One then finds that the relation between the element a of the group SL(2, C) and the element g of the group L are given by[1]

$$g_{\alpha\beta} = (1/2) \, \mathrm{Tr}(\sigma^\alpha a \sigma^\beta a^\dagger) \, . \qquad\qquad (11.5)$$

It thus follows that the group L is homomorphic to the group SL(2, C) such that to every element g of L there correspond two matrices $\mp$ a of SL(2, C) and, conversely, to every $a \in$ SL(2, C) there corresponds some element $g \in$ L. Accordingly, the description of the representations of the group L is equivalent to that of the group SL(2, C); a representation $g \to D_g$ of L is single- or double-valued according to whether or not D_a is equal to D_{-a} or not.

§ 11.2 The Spinor Representation of the Group SL(2, C)

We now construct the spinor representation which contains all the irreducible finite-dimensional representations of the group SL(2, C). A detailed account of this representation will be given in the Appendix.

We denote by P_{mn} the aggregate of all polynomials $p(z, \bar{z})$ in the variable z and its complex conjugate $\bar{z}$ of degree not exceeding m in z and n in $\bar{z}$, where m and n are fixed non-negative integers

determining the representation. The space P_{mn} is a linear vector space

where the operation of addition and multiplication by a number are

defined in the usual way for polynomials.

An element of the group $SL(2, C)$ will now be denoted by

$$g = \begin{pmatrix} a & b \\ c & d \end{pmatrix} , \qquad (11.6)$$

where $a, b, c,$ and d are complex numbers satisfying the condition

$$ad - bc = 1 .$$

Define the operator D_g in the space P_{mn} by

$$D_g \, p(z, \bar{z}) = (bz + d)^m \, (\bar{b} \, \bar{z} + \bar{d})^n \, p\left(\frac{az+c}{bz+d} , \; \frac{\bar{a} \, \bar{z} + \bar{c}}{\bar{b} \, \bar{z} + \bar{d}} \right) . \qquad (11.7)$$

The correspondence $g \to D_g$ is a linear representation of the group

$SL(2, C)$ as can be easily verified. This is known as the *spinor*

representation of the group $SL(2, C)$ and is of dimension

$(m + 1) \times (n + 1)$.

In order to relate this representation to the familiar *2-component*

spinors, one realizes it in a somewhat different way.

One considers all systems of numbers

$$\phi_{A_1 \ldots A_m \, \dot{X}_1 \ldots \dot{X}_n} ,$$

symmetrical in both the indices $A_1, \ldots, A_m$ and in $\dot{X}_1, \ldots, \dot{X}_n$

taking the values 0 and 1. The set of all such systems of numbers

provides a linear space, denoted by S_{mn}, of dimension $(m + 1)(n + 1)$.

A one-to-one linear mapping between the spaces P_{mn} and S_{mn} can

easily be established. To each system of numbers $\phi_{A_1 \ldots A_m \dot{X}_1 \ldots \dot{X}_n} \in S_{mn}$
there corresponds the polynomial

$$P(z, \bar{z}) = \sum_{\substack{A_1, \ldots, A_m \\ \dot{X}_1, \ldots, \dot{X}_n}} \phi_{A_1 \ldots A_m \dot{X}_1 \ldots \dot{X}_n} \, z^{A_1 + \ldots + A_m} \, \bar{z}^{\dot{X}_1 + \ldots + \dot{X}_n} \tag{11.8}$$

of degree not exceeding m in z and n in $\bar{z}$, and therefore

$p(z, \bar{z}) \in P_{mn}$. On the other hand every polynomial

$$P(z, \bar{z}) = \sum_{r,s} p_{rs} \, z^r \, \bar{z}^s \tag{11.9}$$

in the space P_{mn} can be written in the form (11.8) if one relates the

ϕ's and p's by means of

$$\phi_{A_1 \ldots A_m \dot{X}_1 \ldots \dot{X}_n} = \frac{1}{m!n!} \, P_{rs} \, ,$$

with $A_1 + \ldots + A_m = r$ and $\dot{X}_1 + \ldots + \dot{X}_n = s$.

A second form of the spinor representation is then obtained if one

applies the polynomials (11.8) to Equation (11.7). One finds

$$D_g \, p(z, \bar{z}) = (a_{10}z + a_{00})^m \, (\bar{a}_{10}\bar{z} + \bar{a}_{00})^n$$

$$\times \sum_{\substack{A_1 \ldots A_m \\ \dot{X}_1 \ldots \dot{X}_n}} \phi_{A_1 \ldots A_m \, \dot{X}_1 \ldots \dot{X}_n} \left(\frac{a_{11}z + a_{01}}{a_{10}z + a_{00}} \right)^{A_1 + \ldots + A_m}$$

$$\times \left(\frac{\bar{a}_{11}\bar{z} + \bar{a}_{01}}{\bar{a}_{10}\bar{z} + \bar{a}_{00}} \right)^{\dot{X}_1 + \ldots + \dot{X}_n}$$

$$= \sum \phi_{A_1 \ldots A_m \, \dot{X}_1 \ldots \dot{X}_n} \, (a_{1A_1} z + a_{0A_1}) \ldots (a_{1A_m} z + a_{0A_m})$$

$$\times (\bar{a}_{1\dot{X}_1} \bar{z} + \bar{a}_{0\dot{X}_1}) \ldots (\bar{a}_{1\dot{X}_n} \bar{z} + \bar{a}_{0\dot{X}_n}) \, . \qquad (11.10)$$

Hence one obtains

$$D_g \, p(z, z) = \sum_{\substack{A_1, \ldots, A_m \\ \dot{X}_1, \ldots, \dot{X}_n}} \phi'_{A_1 \ldots A_m \, \dot{X}_1 \ldots \dot{X}_n} \, z^{A_1 + \ldots + A_m} \, \bar{z}^{\dot{X}_1 + \ldots + \dot{X}_n} \qquad (11.11)$$

where we have used the notation

$$\phi'_{A_1 \ldots A_m \, \dot{X}_1 \ldots \dot{X}_n} = \sum_{\substack{B_1 \ldots B_m \\ \dot{Y}_1 \ldots \dot{Y}_n}} a_{A_1 B_1} \ldots a_{A_m B_m} \, \bar{a}_{\dot{X}_1 \dot{Y}_1} \ldots \bar{a}_{\dot{X}_n \dot{Y}_n} \, \phi_{B_1 \ldots B_m \, \dot{Y}_1 \ldots \dot{Y}_n}$$

$$(11.12)$$

and where $a_{11} = a$, $a_{10} = b$, $a_{01} = c$, and $a_{00} = d$.

The quantity

$$\phi_{A_1 \ldots A_m \dot{X}_1 \ldots \dot{X}_n}$$

is called a *spinor*, symmetric in its m undotted indices and in its n

dotted ones, whereas Equation (11.12) expresses its transformation law

under the matrix a ϵ SL(2, C).[2,3]

§ 11.3 <u>Infinitesimal Operators of the Spinor Representation</u>

We can now find the infinitesimal operators L_+, L_-, L_3, and F_+,

F_-, F_3 of the spinor representation discussed in the last section.

The one-parameter subgroups of SL(2, C), corresponding to the one-

parameter subgroups $a_k(t)$ and $b_k(t)$ of the group L, can easily be

obtained using the formula (11.5).[4] In terms of the infinitesimal

matrices $\hat{a}_r$ and $\hat{b}_r$ of the group SL(2, C) they can be written as

$$\hat{a}_k(t) = \exp(t\,\hat{a}_k), \quad \hat{b}_k(t) = \exp(t\,\hat{b}_k) , \qquad (11.13)$$

where $\hat{a}_k = i\sigma^k/2$ and $\hat{b}_k = \sigma^k/2$, and where σ^k are the Pauli spin

matrices given by Equation (11.3). Using the matrices $\hat{a}_k(t)$ and $\hat{b}_k(t)$

in (11.7), differentiating both sides of the obtained equations with

respect to the variable t, and putting t = 0, gives the expressions

for the operators A_k and B_k, from which one then obtains the

operators L's and F's:

$$L_+ = -\frac{\partial}{\partial z} - \bar{z}^2 \frac{\partial}{\partial \bar{z}} + n\bar{z} ,$$

$$L_- = z^2 \frac{\partial}{\partial z} + \frac{\partial}{\partial \bar{z}} - mz ,$$

$$L_3 = -z \frac{\partial}{\partial z} + \bar{z} \frac{\partial}{\partial \bar{z}} + \frac{1}{2}(m - n) ,$$

$$\qquad (11.14)$$

$$F_{+} = i \left(\frac{\partial}{\partial z} - \bar{z}^2 \frac{\partial}{\partial \bar{z}} + n\bar{z} \right) ,$$

$$F_{-} = i \left(-z^2 \frac{\partial}{\partial z} + \frac{\partial}{\partial \bar{z}} + mz \right) ,$$

$$F_{3} = i \left(z \frac{\partial}{\partial z} + \bar{z} \frac{\partial}{\partial \bar{z}} - \frac{1}{2} (m + n) \right) .^{[5]}$$

NOTES AND REFERENCES

1. Compare the analogous formulas for the rotation group given by Equations (2.10) and (2.11) in M. Carmeli, *J. Math. Phys.* $\underline{9}$, 1987 (1968). Equations (11.5) can easily be proved by finding the value of the expression $(1/2) \, \mathrm{Tr}(\sigma^\alpha a \sigma^\beta a^\dagger) x_\beta = (1/2) \, \mathrm{Tr}(\sigma^\alpha a \sigma^\beta x_\beta a^\dagger) = (1/2) \, \mathrm{Tr}(\sigma^\alpha a Q a^\dagger) = (1/2) \, \mathrm{Tr}(\sigma^\alpha Q') = (1/2) \, \mathrm{Tr}(\sigma^\alpha \sigma^\beta x'_\beta) = (1/2) \, \mathrm{Tr}(\sigma^\alpha \sigma^\beta) x'_\beta = \delta^{\alpha\beta} x'_\beta = x'_\alpha = g_{\alpha\beta} x_\beta.$

2. Spinors were first discovered by E. Cartan, *Bull. Soc. Math. France* $\underline{41}$, 53 (1913); *Leçons sur la théorie des spineurs*, Hermann, Paris, 1938; English translation in: *The theory of spinors*, The M.I.T. Press, Cambridge, Massachusetts, 1966.

3. For applications of spinors in general relativity theory see, for example, F.A.E. Pirani, *Lectures on General Relativity*, Prentice-Hall, Inc., Englewood Cliffs, New Jersey, 1965.

4. These matrices for the group $SL(2, C)$ are given by

$$\hat{a}_1(t) = \begin{pmatrix} \cos \frac{t}{2} & i\sin \frac{t}{2} \\ i\sin \frac{t}{2} & \cos \frac{t}{2} \end{pmatrix} ,$$

$$\hat{a}_2(t) = \begin{pmatrix} \cos \frac{t}{2} & -\sin \frac{t}{2} \\ \sin \frac{t}{2} & \cos \frac{t}{2} \end{pmatrix},$$

$$\hat{a}_3(t) = \begin{pmatrix} e^{it/2} & 0 \\ 0 & e^{-it/2} \end{pmatrix},$$

and

$$\hat{b}_1(t) = \begin{pmatrix} \cosh \frac{t}{2} & \sinh \frac{t}{2} \\ \sinh \frac{t}{2} & \cosh \frac{t}{2} \end{pmatrix},$$

$$\hat{b}_2(t) = \begin{pmatrix} \cosh \frac{t}{2} & i\sinh \frac{t}{2} \\ -i\sinh \frac{t}{2} & \cosh \frac{t}{2} \end{pmatrix},$$

$$\hat{b}_3(t) = \begin{pmatrix} e^{t/2} & 0 \\ 0 & e^{-t/2} \end{pmatrix}.$$

5. Dirac's equation for particles with spin ½ makes use of four-
    rowed matrices and hence the wave functions are really *four-
    component* spinors having the form

$$\psi = \begin{pmatrix} \psi^1 \\ \psi^2 \\ \psi^3 \\ \psi^4 \end{pmatrix}.$$

[See P.A.M. Dirac, *Quantum Mechanics*, Oxford, 1930]. One can show,
however, that the Dirac 4-component spinor can actually be

constructed from two 2-component spinors, let us say ϕ_A and χ_A, of the type presented in this chapter. The construction of the Dirac spinor from 2-component spinors can be achieved if one writes ψ in the form

$$\psi = \begin{pmatrix} \phi_0 \\ \phi_1 \\ \bar{\chi}^0 \\ \bar{\chi}^1 \end{pmatrix} ,$$

where the indices in χ^A have been raised by means of the Levi-Civita symbol, namely $\chi^A = \epsilon^{AB} \chi_B$, thus $\chi^0 = \chi_1$ and $\chi^1 = -\chi_0$.

APPENDIX

SPINOR REPRESENTATIONS OF THE LORENTZ GROUP

A1. THE GROUP SL(2, C) AND THE LORENTZ GROUP

Consider the set of all 2×2 matrices

$$a = \begin{pmatrix} a & b \\ c & d \end{pmatrix} , \tag{A.1}$$

where a, b, c, d, are complex numbers which satisfy

$$\det a = ad - bc = 1 . \tag{A.2}$$

This set of matrices forms a group, called the *special linear group of order two* and is denoted by SL(2, C).

The group SL(2, C) is closely related to the Lorentz group L. The relationship is derived as follows:[1,2]

Let us associate with each four-vector

$$x = (x_1, x_2, x_3, x_4)$$

a Hermitean matrix

$$Q = \begin{pmatrix} x_4 + x_3 & x_1 + ix_2 \\ x_1 - ix_2 & x_4 - x_3 \end{pmatrix} . \qquad (A.3)$$

This defines a one-to-one linear correspondence between all the four-vectors and all the 2×2 Hermitean matrices. Equation (A.3) can be also formulated as

$$Q = x_\alpha \sigma^\alpha , \qquad (A.4)$$

where σ^k, k = 1, 2, 3, are the Pauli matrices

$$\sigma^1 = \begin{pmatrix} 0 & 1 \\ 1 & 0 \end{pmatrix} , \quad \sigma^2 = \begin{pmatrix} 0 & i \\ -i & 0 \end{pmatrix} , \quad \sigma^3 = \begin{pmatrix} 1 & 0 \\ 0 & -1 \end{pmatrix} , \qquad (A.5)$$

and σ^4 is the 2×2 unit matrix

Corresponding to any given element $a \in SL(2, C)$, consider the following transformation in the space of the Hermitean matrices Q:

$$Q' = a\, Q\, a^\dagger \qquad (A.6)$$

where $a^\dagger$ is the Hermitean conjugate of a and

$$Q' = x'_\alpha \sigma^\alpha . \qquad (A.7)$$

Equations (A.4), (A.6) and (A.7) define a linear transformation in the space of the four-vectors $x = (x_1, x_2, x_3, x_4)$,

$$x'_\alpha = g_{\alpha\beta}\, x_\beta , \qquad (A.8)$$

or, in matrix notation,

$$x' = g\, x\ ,\qquad\qquad (A.9)$$

where the matrix elements $g_{\alpha\beta}$ can be expressed in terms of the matrix elements of a, as follows:

$$x'_\alpha = \delta_{\alpha\beta}\, x'_\beta$$

$$= \tfrac{1}{2}\, \mathrm{Tr}\ (\sigma^\alpha\, \sigma^\beta)\, x'_\beta$$

$$= \tfrac{1}{2}\, \mathrm{Tr}\ (\sigma^\alpha\, \sigma^\beta\, x'_\beta)$$

$$= \tfrac{1}{2}\, \mathrm{Tr}\ (\sigma^\alpha\, Q')$$

$$= \tfrac{1}{2}\, \mathrm{Tr}\ (\sigma^\alpha\, aQa^\dagger)$$

$$= \tfrac{1}{2}\, \mathrm{Tr}\ (\sigma^\alpha a\, \sigma^\beta\, x_\beta a^\dagger)$$

$$= \tfrac{1}{2}\, \mathrm{Tr}\ (\sigma^\alpha a\, \sigma^\beta a^\dagger)\, x_\beta\ .\qquad\qquad (A.10)$$

Therefore,

$$g_{\alpha\beta} = \tfrac{1}{2}\, \mathrm{Tr}\ (\sigma^\alpha a\, \sigma^\beta a^\dagger)\ .\qquad\qquad (A.11)$$

Corresponding to an arbitrary matrix $a \in SL(2, C)$ there exists, therefore, a 4×4 matrix g, defining a linear transformation in ordinary space-time, whose matrix elements are given by Eq. (A.11). Let us now establish that the matrix g belongs to the proper orthochronous Lorentz group L.

(a)   From Eq. (A.3) one has

$$\det Q = x_4{}^2 - x_1{}^2 - x_2{}^2 - x_3{}^2\ ,\qquad\qquad (A.12)$$

and from Eq. (A.6), because of (A.2), it follows that

$$\det Q' = \det Q \; , \qquad\qquad\qquad (A.13)$$

i.e. the transformation (A.11) leaves the quadratic form

$$x_4^{\;2} - x_1^{\;2} - x_2^{\;2} - x_3^{\;2}$$

unchanged, and therefore the matrix g is an element of the complete Lorentz group.

(b) Since g belongs to the complete Lorentz group, it satisfies

$$\det g = \pm 1 \; . \qquad\qquad\qquad (A.14)$$

For the special case

$$a = \begin{pmatrix} 1 & 0 \\ 0 & 1 \end{pmatrix} \; ,$$

g is the identity transformation and $\det g = 1$. Since $\det g$ is a *continuous* function of the four variables a, b, c, d and the domain of variation of these four variables is simply connected (they can take all complex values consistent with Eq. (A.2)), a discontinuous jump from $\det g = 1$ to $\det g = -1$ is excluded, and $\det g = +1$ for all values of a, b, c, d subject to the restriction (A.2); i.e. g belongs to the *proper* Lorentz group.

(c) By Eq. (A.11)

$$g_{44} = \frac{1}{2} \left(|a|^2 + |b|^2 + |c|^2 + |d|^2 \right) > 0 , \qquad (A.15)$$

and therefore g is *orthochronous*, as well as proper, Lorentz transformation.

Let us show now that the correspondence we have established preserves the group multiplication. Let $\pm a \, \epsilon \, SL(2, C)$ correspond to a given element $g \, \epsilon \, L$, and $\pm a' \, \epsilon \, SL(2, C)$ correspond to another element $g' \, \epsilon \, L$. Let us prove that $\pm a'' = \pm a'a \, \epsilon \, SL(2, C)$ will correspond to $g'' = g'g \, \epsilon \, L$.

By Eq. (A.2)

$$x'' = g'(g\,x) = (g'\,g)x = g''\,x , \qquad (A.16)$$

and by Eq. (A.6)

$$Q'' = a'(a\,Q\,a^{\dagger})a'^{\dagger}$$
$$= (a'a)\,Q(a^{\dagger}\,a'^{\dagger})$$
$$= (a'a)\,Q(a'a)^{\dagger}$$
$$= a''\,Q\,a''^{\dagger} . \qquad (A.17)$$

Comparison of Eqs. (A.16), (A.17) with Eqs. (A.6), (A.8) shows that a'' is related to g'' in the same way that a is related to g; i.e.,

$$g''_{\alpha\beta} = \frac{1}{2} \operatorname{Tr} (\sigma^{\alpha}\,a''\,\sigma^{\beta}\,a''^{\dagger}) . \qquad (A.18)$$
$$\text{Q.E.D.}$$

Substituting Eqs. (A.1) and (A.5) in (A.11), the matrix g can be explicitly calculated to yield

$$
g =
\begin{pmatrix}
\mathrm{Re}(a\bar{d} + b\bar{c}) & \mathrm{Im}(\bar{a}d + b\bar{c}) & \mathrm{Re}(a\bar{c} - b\bar{d}) & \mathrm{Re}(a\bar{c} + b\bar{d}) \\[2ex]
\mathrm{Im}(a\bar{d} + b\bar{c}) & \mathrm{Re}(\bar{a}d - b\bar{c}) & \mathrm{Im}(a\bar{c} - b\bar{d}) & \mathrm{Im}(a\bar{c} + b\bar{d}) \\[2ex]
\mathrm{Re}(a\bar{b} - c\bar{d}) & \mathrm{Im}(\bar{a}b + c\bar{d}) & \tfrac{1}{2}(a\bar{a} - b\bar{b} - c\bar{c} + d\bar{d}) & \tfrac{1}{2}(a\bar{a} + b\bar{b} - c\bar{c} - d\bar{d}) \\[2ex]
\mathrm{Re}(a\bar{b} + c\bar{d}) & \mathrm{Im}(\bar{a}b - c\bar{d}) & \tfrac{1}{2}(a\bar{a} - b\bar{b} + c\bar{c} - d\bar{d}) & \tfrac{1}{2}(a\bar{a} + b\bar{b} + c\bar{c} + d\bar{d})
\end{pmatrix}.
$$

$$\text{(A.19)}$$

Given an arbitrary matrix $a \in SL(2, C)$, a corresponding matrix $g \in L$ is uniquely and explicitly given by Eq. (A.19).

Given an arbitrary matrix $g \in L$ we will prove that there exist two matrices $\pm a \in SL(2, C)$ such that Eq. (A.11) [or (A.19)] is satisfied.

Consider first the special case of a matrix g corresponding to a Lorentz transformation (boost) in the positive x-direction. Hence one has

$$
g = b_1(t) =
\begin{pmatrix}
\cosh t & 0 & 0 & \sinh t \\
0 & 0 & 0 & 0 \\
0 & 0 & 0 & 0 \\
\sinh t & 0 & 0 & \cosh t
\end{pmatrix}. \qquad \text{(A.20)}
$$

Equating Eq. (A.20) with (A.19) and solving for a, b, c, d one obtains

$$\hat{b}_1(t) = \pm \begin{pmatrix} \cosh \dfrac{t}{2} & \sinh \dfrac{t}{2} \\[2ex] \sinh \dfrac{t}{2} & \cosh \dfrac{t}{2} \end{pmatrix} . \tag{A.21a}$$

Now, any element $g \, \epsilon \, L$ can be expressed in the form

$$g = v \, b_1(t) v' , \tag{A.22}$$

where v, v' are elements of the group 0_3. We have established in Chapter IV a one-to-two correspondence between all the elements of the rotation group 0_3 and all the elements of the group SU_2, which is a sub-group of the group SL(2, C). Because of Eq. (A.6) it follows that the relationship between the matrices u, u' of the group SU_2 and the corresponding matrices v, v' of the group 0_3 is precisely the same as the correspondence between the elements of the group SL(2, C) and the elements of the group L defined by Eq. (A.11). Equation (A.6) is, indeed, a special case of (A.11). By virtue of Eq. (A.22) the given matrix $g \, \epsilon \, L$ corresponds to

$$(\pm u)(\pm \hat{b}_1(t))(\pm u') = \pm (u \, \hat{b}_1(t) u') \, \epsilon \, SL(2, C) .$$

We have established, therefore, the following result:

Theorem: *There exists a two-to-one mapping between all the elements of the group SL(2, C) and all the elements of the orthochronous, proper, Lorentz group L such that to each element of L there correspond two elements of SL(2, C), differing in signs. The mapping conserves the group multiplication and constitutes, therefore, a homomorphism of the group SL(2, C) on the group L.*

The matrix $\hat{b}_1(t)$, corresponding to a Lorentz transformation in the x-direction, is given by Eq. (A.22). The matrices $\hat{b}_2(t)$, $\hat{b}_3(t)$, corresponding to Lorentz transformations in the y and z directions, are easily found to be given by

$$\hat{b}_2(t) = \begin{pmatrix} \cosh\frac{t}{2} & i\ \sinh\frac{t}{2} \\ -i\ \sinh\frac{t}{2} & \cosh\frac{t}{2} \end{pmatrix}, \qquad (A.21b)$$

$$\hat{b}_3(t) = \begin{pmatrix} e^{\frac{t}{2}} & 0 \\ 0 & e^{-\frac{t}{2}} \end{pmatrix}, \qquad (A.21c)$$

and the matrices $\hat{a}_1(t)$, $\hat{a}_2(t)$ and $\hat{a}_3(t)$, corresponding to rotations around the x, y, and z axes, respectively, are given by

$$\hat{a}_1(t) = \begin{pmatrix} \cos\frac{t}{2} & i\ \sin\frac{t}{2} \\ i\ \sin\frac{t}{2} & \cos\frac{t}{2} \end{pmatrix}, \qquad (A.23a)$$

$$\hat{a}_2(t) = \begin{pmatrix} \cos\frac{t}{2} & -\sin\frac{t}{2} \\ \sin\frac{t}{2} & \cos\frac{t}{2} \end{pmatrix}, \qquad (A.23b)$$

and

$$\hat{a}_3(t) = \begin{pmatrix} e^{\frac{it}{2}} & 0 \\ 0 & e^{-\frac{it}{2}} \end{pmatrix}. \qquad (A.23c)$$

Equations (A.21) and (A.23) can be concisely written as

$$\hat{a}_k(t) = \exp\left(\frac{i\sigma^k}{2}\, t\right),$$

$$\tag{A.24}$$

$$\hat{b}_k(t) = \exp\left(\frac{\sigma}{2}\, t\right),$$

where the σ^k are the Pauli spin matrices (A.5).

NOTES AND REFERENCES

1. M. Carmeli and S. Malin, Finite- and Infinite-Dimensional Representations of the Lorentz Group, *Fortschritte der Physik* <u>21</u>, 397 (1973).

2. M.A. Naimark, *Linear Representations of the Lorentz Group*, Pergamon Press, New York, 1964.

A2. SPINOR REPRESENTATIONS OF THE GROUP SL(2, C)

Consider the space of all the pairs of complex numbers (ϕ_1, ϕ_0). The group SL(2, C) is a group of linear transformation in this two-dimensional complex space; an element

$$g = \begin{pmatrix} a & b \\ c & d \end{pmatrix} ,$$

of the group SL(2, C), transforms the pair (ϕ_1, ϕ_0) into (ϕ'_1, ϕ'_0) according to the formula

$$\phi'_1 = a\phi_1 + b\phi_0 ,$$

$$\phi'_0 = c\phi_1 + d\phi_0 ,$$
(A.25a)

or, in a more concise notation

$$\phi'_A = a_{AB}\, \phi_B ; \quad A, B = 1, 0 ,$$
(A.25b)

where the notation has been used according to which $a_{11} = a$, $a_{10} = b$, $a_{01} = c$, $a_{00} = d$, and the summation convention has also been used and according to which repeated capital letters mean summation over the values 1, 0.

For example, Eqs. (A.25) mean

$$\phi'_A = \sum_{B=0}^{1} a_{AB}\,\phi_B \; .\tag{A.26}$$

The quantities (ϕ_1, ϕ_0) are called *elementary two-component spinors*.

It follows from Eq. (A.24) that the complex conjugate of the elementary two component spinors transform according to the formula

$$\bar{\phi}'_1 = \bar{a}\,\bar{\phi}_1 + \bar{b}\,\bar{\phi}_0 \; ,$$
$$\bar{\phi}'_0 = \bar{c}\,\bar{\phi}_1 + \bar{d}\,\bar{\phi}_0 \; ,\tag{A.27}$$

and hence introducing the notation

$$\bar{\phi}_1 = \phi_{\dot{1}} \; , \quad \bar{\phi}_0 = \phi_{\dot{0}} \; ,\tag{A.28}$$

one obtains the law of transformation

$$\phi'_{\dot{A}} = \bar{a}_{\dot{A}\dot{B}}\,\phi_{\dot{B}} \; .\tag{A.29}$$

The quantities $(\phi_{\dot{1}}, \phi_{\dot{0}})$, which transform according to Eq. (A.29), are called *dotted elementary two-component spinors*.

Consider now the space S_{mn} of all systems of complex numbers

$$\phi_{A_1 \ldots A_m \dot{X}_1 \ldots \dot{X}_n} \; ,$$

symmetrical in the indices $A_1, \ldots, A_m$ and in the indices $\dot{X}_1, \ldots, \dot{X}_n$, all indices taking the values 0 and 1. This space is $(m+1)(n+1)$ dimensional. The law of transformation in this space is

$$\phi'_{A_1 \ldots A_m \dot{X}_1 \ldots \dot{X}_n} = a_{A_1 B_1} \cdots a_{A_m B_m} \, \bar{a}_{\dot{X}_1 \dot{Y}_1} \cdots \bar{a}_{\dot{X}_n \dot{Y}_n} \, \phi_{B_1 \ldots B_m \dot{Y}_1 \ldots \dot{Y}_n} . \qquad (A.30)$$

This space is a direct product of m spaces of the (undotted) elementary two-component spinors and n dotted elementary spinors. The quantities

$$\phi_{A_1 \ldots A_m \dot{X}_1 \ldots \dot{X}_n}$$

are called *two-component spinors.*[1]

The linear transformation given by the coefficients

$$a_{A_1 B_1} \cdots a_{A_m B_m} \, \bar{a}_{\dot{X}_1 \dot{Y}_1} \cdots \bar{a}_{\dot{X}_n \dot{Y}_n}$$

is an $(m + 1) \times (n + 1)$ dimensional representation of the group SL(2, C). This is the *matrix form* of the spinor representation. The representation is denoted by

$$D^{\left(\frac{m}{2}, \, \frac{n}{2}\right)} .$$

Example: The representation $D^{\left(\frac{1}{2}, \, \frac{1}{2}\right)}$ is a linear transformation of all systems of complex numbers

$$(\phi_{1\dot{1}}, \, \phi_{1\dot{0}}, \, \phi_{0\dot{1}}, \, \phi_{0\dot{0}}) .$$

The linear transformation, according to Eq. (A.30), is given by

$$\phi'_{A\dot{X}} = a_{AB} \, \bar{a}_{\dot{X}\dot{Y}} \, \phi_{B\dot{Y}} , \qquad\qquad (A.31)$$

and, in matrix form, the matrix corresponding to an element

$$g = \begin{pmatrix} a & b \\ c & d \end{pmatrix}$$

of the group SL(2, C) is given by

$$D^{(\frac{1}{2}, \frac{1}{2})}(g) = \begin{pmatrix} a\bar{a} & a\bar{b} & b\bar{a} & b\bar{b} \\ a\bar{c} & a\bar{d} & b\bar{c} & b\bar{d} \\ c\bar{a} & c\bar{b} & d\bar{a} & d\bar{b} \\ c\bar{c} & c\bar{d} & d\bar{c} & d\bar{d} \end{pmatrix} . \qquad (A.32)$$

A second form of the spinor representations is introduced as follows: we denote by P_{mn} the aggregate of all polynomials $p(z, \bar{z})$ in the variable z and its complex conjugate $\bar{z}$ of degree not exceeding m in z and n in $\bar{z}$, where m and n are fixed non-negative integers. The space P_{mn} is a linear vector space where the operations of addition and multiplication are defined in the usual way for polynomials.

Corresponding to an element

$$g = \begin{pmatrix} a & b \\ c & d \end{pmatrix}$$

of the group SL(2, C) one can define the operator D_g in the space P_{mn} by

$$D_g \, p(z, \bar{z}) = (bz + d)^m \, (\bar{b} \, \bar{z} + \bar{d})^n \, p\left(\frac{az+c}{bz+d}, \, \frac{\bar{a} \, \bar{z}+\bar{c}}{\bar{b} \, \bar{z}+\bar{d}} \right) . \qquad (A.33)$$

One can easily show that the correspondence $g \to D_g$ is a linear representation of the group $SL(2, C)$. This is the second form of the spinor representation of dimension $(m + 1)(n + 1)$.

The relationship between the two forms of the spinor representations is derived as follows: a one-to-one linear mapping betwen the spaces P_{mn} and S_{mn} can easily be established. To each system of numbers $\phi_{A_1 \ldots A_m \dot{X}_1 \ldots \dot{X}_n} \in S_{mn}$ there corresponds the polynomial

$$p(z, \bar{z}) = \phi_{A_1 \ldots A_m \dot{X}_1 \ldots \dot{X}_n} \, z^{A_1 + \ldots + A_m} \, \bar{z}^{\dot{X}_1 + \ldots + \dot{X}_n} , \qquad (A.34)$$

(the summation convention for $A_1, \ldots, A_m$, $\dot{X}_1, \ldots, \dot{X}_n$ is used here!). $A_1 + \ldots + A_m = m$ and $\dot{X}_1 + \ldots + \dot{X}_n = n$, and therefore $p(z, \bar{z}) \in P_{mn}$. Conversely, every polynomial

$$p(z, \bar{z}) = \sum_{r=0}^{m} \sum_{s=0}^{n} p_{rs} \, z^r \, \bar{z}^s \qquad (A.35)$$

in the space P_{mn} can be written in the form (A.34) if one relates the ϕ's and p's by means of

$$\phi_{A_1 \ldots A_m \dot{X}_1 \ldots \dot{X}_n} = \frac{1}{m!n!} \, p_{rs} , \qquad (A.36)$$

with $A_1 + \ldots + A_m = r$ and $\dot{X}_1 + \ldots + \dot{X}_n = s$.

Equation (A.33) is now explicitly given as

$$D_g p(z, \bar{z}) = (a_{10}z + a_{00})^m (\bar{a}_{10}\bar{z} + \bar{a}_{00})^n \phi_{A_1 \ldots A_m \dot{X}_1 \ldots \dot{X}_n}$$

$$\times \left(\frac{a_{11}z + a_{01}}{a_{10}z + a_{00}} \right)^{A_1 + \ldots + A_m} \left(\frac{\bar{a}_{11}\bar{z} + \bar{a}_{01}}{\bar{a}_{10}\bar{z} + \bar{a}_{00}} \right)^{\dot{X}_1 + \ldots + \dot{X}_n}$$

$$= \phi_{A_1 \ldots A_m \dot{X}_1 \ldots \dot{X}_n} (a_{1A_1} z + a_{0A_1}) \ldots (a_{1A_m} z + a_{0A_m})$$

$$\times (\bar{a}_{1\dot{X}_1} \bar{z} + \bar{a}_{0\dot{X}_1}) \ldots (\bar{a}_{1\dot{X}_n} \bar{z} + \bar{a}_{0\dot{X}_n})$$

$$= \phi'_{A_1 \ldots A_m \dot{X}_1 \ldots \dot{X}_n} z^{A_1 + \ldots + A_m} \bar{z}^{\dot{X}_1 + \ldots + \dot{X}_n} \qquad (A.37)$$

where the ϕ's are given by Eq. (A.30). As previously was done, we have
put $a_{11} = a$, $a_{10} = b$, $a_{01} = c$, $a_{00} = d$, where a, b, c, and d are
defined by

$$g = \begin{pmatrix} a & b \\ c & d \end{pmatrix} ,$$

and where g is an element of the group $SL(2, C)$.

Example: The spinor representation corresponding to
$m = n = 1$ is given on the space of polynomials

$$p(z, \bar{z}) = P_{00} + P_{10}z + P_{01}\bar{z} + P_{11}z\bar{z} . \qquad (A.38)$$

The operator D_g corresponding to an element

$$g = \begin{pmatrix} a & b \\ c & d \end{pmatrix}$$

of the group $SL(2, C)$ is given by

$$D_g \, p(z, \bar{z}) = (bz + d)(\bar{b}\,\bar{z} + \bar{d}) \left[P_{00} + P_{10} \frac{az + c}{bz + d} \right.$$

$$\left. + P_{01} \frac{\bar{a}\,\bar{z} + \bar{c}}{\bar{b}\,\bar{z} + \bar{d}} + P_{11} \left(\frac{az + c}{bz + d} \times \frac{\bar{a}\,\bar{z} + \bar{c}}{\bar{b}\,\bar{z} + \bar{d}} \right) \right]$$

$$= (bz + d)(\bar{b}\bar{z} + \bar{d})p_{00} + (az + c)(\bar{b}\bar{z} + \bar{d})p_{10}$$

$$+ (bz + d)(\bar{a}\bar{z} + \bar{c})p_{01} + (az + c)(\bar{a}\bar{z} + \bar{c})p_{11}$$

$$= p'_{00} + p'_{10}\, z + p'_{01}\, \bar{z} + p'_{11}\, z\bar{z} , \qquad \text{(A.39)}$$

where

$$\begin{pmatrix} p'_{11} \\ p'_{10} \\ p'_{01} \\ p'_{00} \end{pmatrix} = \begin{pmatrix} a\bar{a} & a\bar{b} & b\bar{a} & b\bar{b} \\ a\bar{c} & a\bar{d} & b\bar{c} & b\bar{d} \\ c\bar{a} & c\bar{b} & d\bar{a} & d\bar{b} \\ c\bar{c} & c\bar{d} & d\bar{c} & d\bar{d} \end{pmatrix} \begin{pmatrix} p_{11} \\ p_{10} \\ p_{01} \\ p_{00} \end{pmatrix} . \qquad \text{(A.40)}$$

Let us establish now a third form for the spinor representation. To this end we proceed as follows.

Starting from Eq. (A.35) let us denote $p(z, \bar{z})$ by $p(z)$ and let

$$\alpha(g) = g_{22}^{m} \, \bar{g}_{22}^{-n} \, , \tag{A.41}$$

where

$$g = \begin{pmatrix} g_{11} & g_{12} \\ g_{21} & g_{22} \end{pmatrix} \tag{A.42}$$

is an element of the group SL(2, C). Equation (A.33) can then be rewritten in the form

$$D_g \, p(z) = \alpha(zg) \, p(z(g)) \, . \tag{A.43}$$

Here z denotes a complex variable and also the matrix defined by

$$z = \begin{pmatrix} 1 & 0 \\ z & 1 \end{pmatrix} \, , \tag{A.44}$$

and the matrix $z' = z(g)$ amounts to a transformation in which the variable z goes over into the new variable

$$z' = g'_{21}/g'_{22} \, , \tag{A.45}$$

where the matrix $g' \in$ SL(2, C) is given by

$$g' = zg = \begin{pmatrix} g_{11} & g_{12} \\ g_{11}z + g_{21} & g_{12}z + g_{22} \end{pmatrix} \, , \tag{A.46}$$

so that the new variable z', according to Eqs. (A.45) and (A.46), is given by

$$z' = \frac{g_{11}z + g_{21}}{g_{12}z + g_{22}} \, . \qquad (A.47)$$

The third form of the spinor representation will be obtained by expressing Eq. (A.43) in terms of matrix elements of the group SU_2. Let

$$u = \begin{pmatrix} u_{11} & u_{12} \\ u_{21} & u_{22} \end{pmatrix} \qquad (A.48)$$

denotes an element of SU_2. Let $\tilde{P}_{mn}$ denote the space of all polynomials $q(u)$ which are homogeneous in u_{21}, u_{22} of degree m and in $\bar{u}_{21}$, $\bar{u}_{22}$ of degree n and which also satisfy the condition

$$q(\gamma u) = e^{i(m-n)\psi/2} \, q(u) \, , \qquad (A.49)$$

where the matrix $\gamma \in SU_2$ is given by

$$\gamma = \begin{pmatrix} e^{-i\psi/2} & 0 \\ 0 & e^{i\psi/2} \end{pmatrix} \, . \qquad (A.49a)$$

We proceed to establish a mapping between the space $\tilde{P}_{mn}$ and the space P_{mn} of polynomials $p(z)$.

Definition: The set of all matrices k having the form

$$k = \begin{pmatrix} \lambda^{-1} & \mu \\ 0 & \lambda \end{pmatrix} \, , \qquad (A.50)$$

with λ, μ complex numbers and $\lambda \neq 0$, forms a subgroup of SL(2, C).

This subgroup will be denoted by K.

Definition: The set of all matrices z having the form

$$z = \begin{pmatrix} 1 & 0 \\ z & 1 \end{pmatrix} \, ,$$

(A.51)

with z a complex number, forms a subgroup of SL(2, C). It will be
denoted by Z. (The same notation is used for the matrix z ε Z and
the complex number z. It will be always clear from the context which
is meant).

Lemma: *Any element g ε SL(2, C) satisfying $g_{22} \neq 0$ can be
uniquely decomposed in the form*

$$g = k \, z \, , \quad k \, \varepsilon \, K \, , \quad z \, \varepsilon \, Z \, .$$

(A.52)

Proof: Given a matrix g ε SL(2, C), i.e.

$$g = \begin{pmatrix} g_{11} & g_{12} \\ g_{21} & g_{22} \end{pmatrix} \, ,$$

(A.53)

$$g_{11} \, g_{22} - g_{12} \, g_{21} = 1 \, ,$$

(A.54)

equation (A.52) reads

$$\begin{pmatrix} g_{11} & g_{12} \\ g_{21} & g_{22} \end{pmatrix} = \begin{pmatrix} \lambda^{-1} & \mu \\ 0 & \lambda \end{pmatrix} \begin{pmatrix} 1 & 0 \\ z & 1 \end{pmatrix} \, .$$

(A.55)

A straightforward calculation shows that because of Eq. (A.54), it follows that Eq. (A.55) has the unique solution

$$\lambda = g_{22} , \quad \mu = g_{12} ,$$
$$\text{(A.56)}$$

$$z = g_{21}/g_{22} .$$
$$\text{(A.57)}$$

Consider now the set of all right cosets (see Chapter I) of the group SL(2, C) with respect to the group K. It follows from the lemma that there exists a one-to-one correspondence between all these right cosets and all the complex numbers z: because of Eq. (A.57) all the elements of the group SL(2, C) for which the ratio $g_{12}/g_{22} = z$ is the same, belong to the same right coset, which will be denoted by $\tilde{z}$.

It will now be proved that every right coset $\tilde{z}$ of the group SL(2, C) with respect to the group K contains elements of the group SU_2, which is a subgroup of SL(2, C).

Lemma: *Every element $g \in SL(2, C)$ can be decomposed in the form*

$$g = k\,u , \quad k \in K , \quad u \in SU_2 .$$
$$\text{(A.58)}$$

Proof: The general form of the elements $u \in SU_2$ is given by

$$u = \begin{pmatrix} \alpha & \beta \\ -\bar{\beta} & \bar{\alpha} \end{pmatrix} ,$$
$$\text{(A.59)}$$

with α, β being complex numbers satisfying the condition

$$|\alpha|^2 + |\beta|^2 = 1 \; . \tag{A.60}$$

Equation (A.58) can be written in the form

$$\begin{pmatrix} g_{11} & g_{12} \\ g_{21} & g_{22} \end{pmatrix} = \begin{pmatrix} \lambda^{-1} & \mu \\ 0 & \lambda \end{pmatrix} \begin{pmatrix} \alpha & \beta \\ -\bar{\beta} & \bar{\alpha} \end{pmatrix} \; . \tag{A.61}$$

Solving Eq. (A.61) for the variables λ, μ, α, β one obtains, because of Eq. (A.60),

$$|\lambda|^2 = |g_{21}|^2 + |g_{22}|^2 \; , \tag{A.62}$$

$$\alpha = \frac{\bar{g}_{21}}{\bar{\lambda}} \; , \tag{A.63}$$

$$\beta = -\frac{\bar{g}_{21}}{\bar{\lambda}} \; , \tag{A.64}$$

$$\mu = \begin{cases} \dfrac{g_{12} - \beta/\lambda}{\bar{\alpha}} \; , & \text{when } \alpha \neq 0 \; , \\[2em] -\dfrac{g_{11}}{\bar{\beta}} \; , & \text{when } \alpha = 0 \; . \end{cases} \tag{A.65}$$

$$\text{Q.E.D.}$$

It follows, therefore, that the decomposition (A.52) holds for an arbitrary element $g \in SL(2, C)$, and that, furthermore, the decomposition is not unique, because the phase of λ is left undetermined by Eqs. (A.62)-(A.65). In fact, if γ is any matrix of the form

$$\gamma = \begin{pmatrix} e^{-i\omega} & 0 \\ 0 & e^{i\omega} \end{pmatrix} , \quad \omega \text{ real} \tag{A.66}$$

then $\gamma \in SU_2$, and if $k \in K$ then $k \gamma \in K$. Therefore

$$g = k \ u = (k \ \gamma)(\gamma^{-1} \ u) = k' \ u' , \tag{A.67}$$

with $k, k' \in K$ and $u, u' \in SU_2$, with γ being arbitrary element of the form (A.66).

From Eqs. (A.52), (A.58) and (A.67) it follows that each right coset $\check{z}$ of the group SL(2, C) with respect to the group K contains a one-parametric set of elements belonging to the group SU_2. Because of Eq. (A.57) it follows that all of the matrices of SU_2 belonging to a given coset $\check{z}$ satisfy

$$- \frac{\bar{\beta}}{\bar{\alpha}} = \frac{u_{21}}{u_{22}} = z . \tag{A.68}$$

We are now in a position to establish the third form of the spinor representation.[2]

Let the polynomial $q(u)$ be defined as

$$q(u) = \pi^{\frac{1}{2}} \alpha(u) \ p(z) , \tag{A.69}$$

where, by Eq. (A.41),

$$\alpha(u) = u_{22}^{\ m} \ \bar{u}_{22}^{\ n} , \tag{A.70}$$

and where the matrices   u   and the matrix

$$z = \begin{pmatrix} 1 & 0 \\ z & 1 \end{pmatrix}$$

belong to the same right coset $\tilde{z}$.   Since $z = u_{21}/u_{22}$ by Eq. (A.68),
one has

$$q(u) = \pi^{\frac{1}{2}} \sum_{r=0}^{m} \sum_{s=0}^{n} P_{rs} \, u_{21}^{r} \, u_{22}^{m-r} \, \bar{u}_{21}^{-s} \, \bar{u}_{22}^{-n-s} \; . \qquad (A.71)$$

The polynomial $q(u)$ satisfies the condition

$$q(\gamma u) = e^{i(m-n)\psi/2} \, q(u)$$

because, by Eq. (A.49a), $(\gamma u)_{21} = u_{21}$, $(\gamma u)_{22} = e^{i\psi/2} \, u_{22}$;   therefore
the polynomial $q(u)$ belongs to the space $\tilde{P}_{mn}$.   Consequently,
Eq. (A.69) defines a mapping between the space P_{mn} of the polynomials
$p(z)$ and the space $\tilde{P}_{mn}$ of the polynomials $q(u)$.

By Eqs. (A.43) and (A.69) the operators of the spinor
representations in the space $\tilde{P}_{mn}$ are given by the formula

$$D_{g} \, q(u) = \frac{\alpha(ug)}{\alpha(u(g))} \, q(u(g)) \; , \qquad (A.72)$$

where $u(g)$ is a matrix of the group SU_2 which belong to the right
coset $zg = \tilde{z}'$, where

$$z' = \frac{g_{11}z + g_{21}}{g_{12}z + g_{22}} \; . \qquad (A.73)$$

The matrix $u(g)$ can be explicitly obtained in terms of the matrices $u \in SU_2$ and $g \in SL(2, C)$ as follows:[3,4]

Let us denote the matrix $u(g)$ by u'. Then u' can be written as

$$u' = \begin{pmatrix} u'_{11} & u'_{12} \\ u'_{21} & u'_{22} \end{pmatrix} = \begin{pmatrix} \alpha & \beta \\ -\bar{\beta}' & \bar{\alpha}' \end{pmatrix} \qquad (A.74)$$

along with the condition

$$|\alpha'|^2 + |\beta'|^2 = 1 . \qquad (A.75)$$

According to Eq. (A.58) ug can be written in the form $ug = k\,u\,\bar{g} = ku'$, where k is a matrix having the form given by Eq. (A.50). If one denotes now ug by g', then one has $g' = ku'$, or explicitly

$$\begin{pmatrix} g'_{11} & g'_{12} \\ g'_{21} & g'_{22} \end{pmatrix} = \begin{pmatrix} \lambda^{-1} & \mu \\ 0 & \lambda \end{pmatrix} \begin{pmatrix} \alpha' & \beta' \\ -\bar{\beta}' & \bar{\alpha}' \end{pmatrix} . \qquad (A.76)$$

This gives

$$g'_{21} = -\lambda\,\bar{\beta}' , \quad g'_{22} = \lambda\,\bar{\alpha}' , \qquad (A.77)$$

from which one obtains

$$\alpha' = \frac{g'_{22}}{\bar{\lambda}} , \quad \beta' = -\frac{\bar{g}'_{21}}{\bar{\lambda}} . \qquad (A.78)$$

Furthermore, using the condition (A.75), one obtains

$$|\lambda|^2 = |g'_{21}|^2 + |g'_{22}|^2 \, .$$

(A.79)

But $g' = ug$. Let us denote u by

$$u = \begin{pmatrix} u_{11} & u_{12} \\ u_{21} & u_{22} \end{pmatrix} = \begin{pmatrix} \alpha & \beta \\ -\bar{\beta} & \bar{\alpha} \end{pmatrix} \, ,$$

(A.80)

and g by

$$g = \begin{pmatrix} g_{11} & g_{12} \\ g_{21} & g_{22} \end{pmatrix} \, ,$$

(A.81)

then

$$\begin{pmatrix} g'_{11} & g'_{12} \\ g'_{21} & g'_{22} \end{pmatrix} = \begin{pmatrix} \alpha g_{11} + \beta g_{21} & \alpha g_{12} + \beta g_{22} \\ -\bar{\beta} g_{11} + \bar{\alpha} g_{21} & -\bar{\beta} g_{12} + \bar{\alpha} g_{22} \end{pmatrix} \, .$$

(A.82)

If we write now $\lambda = |\lambda| \exp(i \Lambda)$, where Λ is some real number (phase),
then one finally obtains for Eqs. (A.78) and (A.79)

$$\alpha' = (-\beta \bar{g}_{12} + \alpha \bar{g}_{22}) \, |\lambda|^{-1} \, e^{i\Lambda} \, ,$$

$$\beta' = (\beta \bar{g}_{11} - \alpha \bar{g}_{21}) \, |\lambda|^{-1} \, e^{i\Lambda} \, ,$$

(A.83)

and

$$|\lambda|^2 = |\beta \bar{g}_{11} - \alpha \bar{g}_{21}|^2 + |-\beta \bar{g}_{12} + \alpha \bar{g}_{22}|^2 \, .$$

(A.84)

Hence, $u(g)$ is determined by means of u and g up to an arbitrary phase factor. It is readily verified that the right-hand side of Eq. (A.72) is independent of the arbitrary phase factor, because the phase factor cancels out when the ratio $q(u(g))/\alpha(u(g))$ is calculated. The spinor representations are, therefore, well defined on the space $\tilde{P}_{mn}$ by Eq. (A.72).

Examples:

(1) Let g be a unitary matrix u_0 with determinant unity:

$$u_0 = \begin{pmatrix} \alpha_0 & \beta_0 \\ -\bar{\beta}_0 & \bar{\alpha}_0 \end{pmatrix} \; ; \quad |\alpha_0|^2 + |\beta_0|^2 = 1. \tag{A.85}$$

Then one obtains from Eqs. (A.83), (A.84)

$$\alpha' = (-\beta\,\bar{\beta}_0 + \alpha\alpha_0)\,e^{i\Lambda}\,,$$

$$\beta' = (\beta\,\bar{\alpha}_0 + \alpha\beta_0)\,e^{i\Lambda}\,, \tag{A.86}$$

$$|\lambda| = 1\,,$$

and accordingly

$$\frac{\alpha(uu_0)}{\alpha(u(u_0))} = e^{is\Lambda}\,. \tag{A.87}$$

Here $s = \tfrac{1}{2}(m - n)$.

(2) Let $g \in SL(2, C)$ be of the form

$$g = \begin{pmatrix} \varepsilon_{22}^{-1} & 0 \\ 0 & \varepsilon_{22} \end{pmatrix} , \tag{A.88}$$

where ε_{22} is real and different from zero. One then obtains

$$\alpha' = \alpha\, \varepsilon_{22} |\lambda|^{-1} e^{i\Lambda} ,$$

$$\tag{A.89}$$

$$\beta' = \beta\, \varepsilon_{22}^{-1} |\lambda|^{-1} e^{i\Lambda} ,$$

$$|\lambda|^2 = |\beta|^2 \varepsilon_{22}^{-2} + |\alpha|^2 \varepsilon_{22}^{2} , \tag{A.90}$$

and

$$\frac{\alpha(u\varepsilon)}{\alpha(u(\varepsilon))} = |\lambda|^{i\rho - 2} e^{2is\Lambda} . \tag{A.91}$$

Here one has $s = \tfrac{1}{2}(m - n)$ and $i\rho - 2 = (m + n)$.

NOTES AND REFERENCES

1. E. Cartan, *Leçon sur la théorie des spineurs*, Herman, Paris, 1938;
    English translation in: *The Theory of Spinors*, The M.I.T. Press,
    Cambridge, Massachusetts, 1966.

2. M.A. Naimark, *Linear Representations of the Lorentz Group*,
    Pergamon Press, New York, 1964.

3. M. Carmeli and S. Malin, Infinite-Dimensional Representations of
    the Lorentz Group: Complementary Series of Representations,

J. Math. Phys. 12, 225 (1971).

4. M. Carmeli and S. Malin, Finite- and Infinite-Dimensional
    Representations of the Lorentz Group, *Fortschritte der Physik* 21,
    397 (1973).

A3.   INFINITESIMAL OPERATORS FOR THE SPINOR REPRESENTATIONS

In the present section we find the infinitesimal operators L_+ , L_- , L_3 , and F_+ , F_- , F_3 of the spinor representations.

In the previous chapter the one parameter subgroups $a_k(t)$ and $b_k(t)$ of the group L were introduced. The subgroups $a_1(t)$, $a_2(t)$, $a_3(t)$ were the groups of rotations around the x, y, z axes respectively, and the parameter t was the angle of rotation; the subgroups $b_1(t)$, $b_2(t)$, $b_3(t)$ were the groups of Lorentz transformations along the x, y, z axes respectively and the parameter t was defined by

$$t = \cosh^{-1} \left[(1 - \frac{v^2}{c^2})^{-\frac{1}{2}} \right] ,$$

where v is the relative speed.

The corresponding subgroups $\hat{a}_k(t)$ and $\hat{b}_k(t)$ of the group SL(2, C) were obtained and are given by Eqs. (A.21) and (A.23). Expanding the $\hat{a}_k(t)$ and $\hat{b}_k(t)$ in power series in t one obtains to first order in t:

$$\hat{a}_1(t) \approx \begin{pmatrix} 1 & i\frac{t}{2} \\ i\frac{t}{2} & 1 \end{pmatrix} ,$$

$$\hat{a}_2(t) \simeq \begin{pmatrix} 1 & -\dfrac{t}{2} \\ \dfrac{t}{2} & 1 \end{pmatrix} , \qquad\qquad\qquad \text{(A.92a)}$$

$$\hat{a}_3(t) \simeq \begin{pmatrix} 1 + \dfrac{it}{2} & 0 \\ 0 & 1 - \dfrac{it}{2} \end{pmatrix} ,$$

and

$$\hat{b}_1(t) \simeq \begin{pmatrix} 1 & \dfrac{t}{2} \\ \dfrac{t}{2} & 1 \end{pmatrix} ,$$

$$\hat{b}_2(t) \simeq \begin{pmatrix} 1 & i\dfrac{t}{2} \\ -i\dfrac{t}{2} & 1 \end{pmatrix} , \qquad\qquad\qquad \text{(A.92b)}$$

$$\hat{b}_3(t) \simeq \begin{pmatrix} 1 + \dfrac{t}{2} & 0 \\ 0 & 1 - \dfrac{t}{2} \end{pmatrix} .$$

By Eq. (A.33) the spinor representation of the one-parameter subgroups $\hat{a}_k(t)$ and $\hat{b}_k(t)$ is given to first order in t by

$$\hat{A}_1(t) \, p(z, \bar{z}) \simeq (1 + \frac{im}{2} tz)(1 - \frac{in}{2} \bar{z}) \, p\left(\frac{z + \dfrac{it}{2}}{1 + \dfrac{it}{2} z} , \ \frac{\bar{z} - \dfrac{it}{2}}{1 - \dfrac{it}{2} \bar{z}} \right) ,$$

$$\hat{A}_2(t)\, p(z,\,\bar{z}) \simeq (1 - \tfrac{m}{2}\, tz)(1 - \tfrac{n}{2}\, t\bar{z})\; p\!\left(\frac{z + \tfrac{t}{2}}{1 - \tfrac{t}{2}\, z}\, , \; \frac{\bar{z} + \tfrac{t}{2}}{1 - \tfrac{t}{2}\, \bar{z}} \right) ,$$

$$\hat{A}_3(t)\, p(z,\,\bar{z}) \simeq (1 - \tfrac{im}{2}\, t)(1 + \tfrac{int}{2})\; p(z + itz,\; \bar{z} - it\bar{z}) ,$$

(A.93a)

and

$$\hat{B}_1(t)\, p(z,\,\bar{z}) \simeq (1 + \tfrac{im}{2}\, tz)(1 + \tfrac{in}{2}\, t\bar{z})\; p\!\left(\frac{z + \tfrac{t}{2}}{1 + \tfrac{t}{2}\, z}\, , \; \frac{\bar{z} + \tfrac{t}{2}}{1 + \tfrac{t}{2}\, \bar{z}} \right) ,$$

$$\hat{B}_2(t)\, p(z,\,\bar{z}) \simeq (1 + \tfrac{im}{2}\, tz)(1 - \tfrac{in}{2}\, t\bar{z})\; p\!\left(\frac{z - i\tfrac{t}{2}}{1 + i\tfrac{t}{2}\, z}\, , \; \frac{\bar{z} + i\tfrac{t}{2}}{1 - i\tfrac{t}{2}\, \bar{z}} \right) ,$$

$$\hat{B}_3(t)\, p(z,\,\bar{z}) \simeq (1 - \tfrac{mt}{2})(1 - \tfrac{nt}{2})\; p(z + tz,\; \bar{z} + t\bar{z}) .$$

(A.93b)

The infinitesimal operators $\hat{A}_1$, $\hat{A}_2$, $\hat{A}_3$ and $\hat{B}_1$, $\hat{B}_2$, $\hat{B}_3$ are obtained by differentiating Eqs. (A.93) with respect to t and taking the limit as $t \to 0$. The operators $L_\pm$, L_3 and $F_\pm$, F_3 are then obtained by taking the appropriate linear combinations, namely,

$$L_+\, p(z,\,\bar{z}) = (i\,\hat{A}_1 - \hat{A}_2)\, p(z,\,\bar{z}) = \left(-\frac{\partial}{\partial z} - \bar{z}^2 \frac{\partial}{\partial \bar{z}} + n\bar{z} \right) p(z,\,\bar{z}) ,$$

$$L_-\, p(z,\,\bar{z}) = (i\,\hat{A}_1 + \hat{A}_2)\, p(z,\,\bar{z}) = \left(z^2 \frac{\partial}{\partial z} + \frac{\partial}{\partial \bar{z}} - mz \right) p(z,\,\bar{z}) ,$$

$$L_3 \, p(z, \bar{z}) = i \, \hat{A}_3 \, p(z, \bar{z}) = (-z \, \frac{\partial}{\partial z} + \bar{z} \, \frac{\partial}{\partial \bar{z}} + \frac{1}{2} m - \frac{1}{2} n) \, p(z, \bar{z}) \; ,$$

$$(A.94a)$$

and

$$F_+ \, p(z, \bar{z}) = (i \, \hat{B}_1 - \hat{B}_2) \, p(z, \bar{z}) = (i \, \frac{\partial}{\partial z} - i \, \bar{z}^2 \, \frac{\partial}{\partial \bar{z}} + in \, \bar{z}) \, p(z, \bar{z}) \; ,$$

$$F_- \, p(z, \bar{z}) = (i \, \hat{B}_1 + \hat{B}_2) \, p(z, \bar{z}) = (-i \, z^2 \, \frac{\partial}{\partial z} + i \, \frac{\partial}{\partial \bar{z}} + im \, \bar{z}) \, p(z, \bar{z}) \; ,$$

$$F_3 \, p(z, \bar{z}) = i \, \hat{B}_3 \, p(z, \bar{z}) = (iz \, \frac{\partial}{\partial z} + i\bar{z} \, \frac{\partial}{\partial \bar{z}} + \frac{i}{2} m + \frac{i}{2} n) \, p(z, \bar{z}) \; .$$

$$(A.94b)$$

Let us establish some important properties of the spinor representations:

(1) *All the spinor representations are irreducible.*

Let $p(z, \bar{z}) \, \epsilon \, P_{mn}$ be monomial,

$$p(z, \bar{z}) = z^r \, \bar{z}^s \; , \qquad 0 \leqslant r \leqslant m \; , \quad 0 \leqslant s \leqslant n \; . \qquad (A.95)$$

Hence by Eq. (A.94) one obtains

$$(-L_+ - iF_+)(z^r \, \bar{z}^s) = \frac{\partial}{\partial z} (z^r \, \bar{z}^s) = r z^{r-1} \, \bar{z}^s \; ,$$

$$(L_- - iF_-)(z^r \, \bar{z}^s) = \frac{\partial}{\partial \bar{z}} (z^r \, \bar{z}^s) = s \, z^r \, \bar{z}^{s-1} \; ,$$

$$(A.96)$$

$$(L_1 + iF_-)(z^r \, \bar{z}^s) = (z^r \, \frac{\partial}{\partial z} - mz)(z^r \, \bar{z}^s) = (r - m) z^{r+1} \, \bar{z}^s \; ,$$

$$(-L_+ + iF_+)(z^r \bar{z}^s) = (\bar{z}^r \frac{\partial}{\partial \bar{z}} - n\bar{z})(z^r \bar{z}^s) = (s - n) z^r \bar{z}^{s+1} \ .$$

It follows from Eqs. (A.96), by repeated applications of the operators $\mp L_\pm - iF_\pm$ and $\pm L_{\mp} + iF_{\mp}$, that the space P_{mn} contains, together with the monomial $z^r \bar{z}^s$, all monomials $z^a \bar{z}^b$ with $0 \leqslant a \leqslant m$, $0 \leqslant b \leqslant n$. P_{mn} does not contain any invariant subspace and therefore the spinor representations are irreducible.

>   (2)   *All the finite dimensional irreducible representations of the*
>   *group SL(2, C) are equivalent to spinor representations.*
>
>   The group SL(2, C) has two *Casimir operators:*[1]

$$C_1 = F^2 - L^2 = \frac{1}{2}(F_+F_- + F_-F_+ + F_3^2 - L_+L_- - L_-L_+ - L_3^2) \ ,$$

$$\text{(A.97)}$$

$$C_2 = \vec{L} \cdot \vec{F} + \vec{F} \cdot \vec{L} = L_+F_- + L_-F_+ + F_+L_- + F_-L_+ + 2L_3F_3 \ .$$

Substitution of Eq. (A.94) in Eq. (A.97) yields the values of the Casimir operators for the spinor representations:

$$C_1 \ p(z, \bar{z}) = -\frac{1}{2}(m^2 + n^2 + 2m + 2n) \ p(z, \bar{z}) \ ,$$

$$\text{(A.98)}$$

$$C_2 \ p(z, \bar{z}) = -\frac{i}{2}(m + n + 2)(m - n) \ p(z, \bar{z}) \ .$$

On the other hand, substitution of the algebraic form of the infinitesimal operators in Eq. (A.97) yields

$$C_1 \ f_m^j = (1 - j_0^2 - c^2) f_m^j \ ,$$

$$\text{(A.99)}$$

$$C_2 \, f_m^j = - 2i j_0 c \, f_m^j \, ,$$

and all the finite dimensional representations are characterized by pairs of real numbers (j_0, c) where j_0 integer or half-integer and $c^2 = (j_0 + n)^2$ with n an arbitrary natural number.

Comparing the right hand side of Eqs. (A.98) and (A.99) one obtains

$$-(1/2)(m^2 + n^2 + 2m + 2n) = 1 - j_0^2 - c^2 \, ,$$

$$\text{(A.100)}$$

$$-(i/2)(m + n + 2)(m - n) = - 2i j_0 c \, .$$

Corresponding to any pair (j_0, c), with $2j_0$ an integer and $c^2 = (j_0 + n)^2$ with n a natural number, Eqs. (A.100) have solutions m, n which are natural numbers. Therefore all the finite-dimensional irreducible representations of the group $SL(2, C)$ can be realized on the space of polynomials P_{mn} and are equivalent to the spinor representations.

(3)   *The spinor representations are all non-unitary. The group SL(2, C) does not contain finite-dimensional unitary representations.*

This statement follows from the unitarity conditions derived in Chapter X.

NOTES AND REFERENCES

1. Casimir operators are defined as operators which commute with all the infinitesimal operators of the group. For more details on the Casimir operators see, for example, M. Hamermesh, *Group Theory*, Addison-Wesley Publishing Company, Inc., Reading, Massachusetts, 1962.

LIST OF SYMBOLS

Symbol	*Description*
$a_1(\psi)$, $a_2(\psi)$, $a_3(\psi)$	Rotations
a_1, a_2, a_3	Infinitesimal matrices of rotations
$a^\dagger$	Hermitian conjugate
a	Element of the group $SL(2, C)$
A_k	Basic infinitesimal operators
$b_1(\psi)$, $b_2(\psi)$, $b_3(\psi)$	Lorentz transformations
b_1, b_2, b_3	Infinitesimal matrices of Lorentz transformations
B_k	Basic infinitesimal operators
D_g, D_u	Representation operators
dg	Measure with respect to the group 0_3
du	Measure with respect to the group SU_2
$D^j_{mn}(u)$	Matrix elements of irreducible representations of the group SU_2
$D^{(m,n)}$	Matrix form of spinor representation
$e_1, \ldots, e_n$	Basis in space R

e — Unit element of group

ε_{rst} — Skew-symmetric Levi-Civita tensor

$f_{-j}, \ldots, f_j$ — Canonical basis

$F_{\pm}, F_3$ — Lorentz (boost) operators

ϕ_1, θ, ϕ_2 — Euler angles

$\phi_{A_1 \ldots A_m \dot{X}_1 \ldots \dot{X}_n}$ — Two-component spinor

G — Group

G/N — Factor group

g_k — Infinitesimal matrices

g^t — Transposed of g

γ — Special unitary matrix

H — Subgroup

$L_{\pm}, L_3$ — Angular momentum operators

Λ — Phase

M^j — Subspace

N — Normal subgroup

$\vec{n}$ — Unit vector

O_3 — Three-dimensional pure rotation group

P — Hermitian matrix

P_{mn} — Space of polynomials

$p(z_1, z_2)$ — Polynomial

$p(z)$ — Polynomial

Q	Hermitian matrix
R	Linear space
S_k	Subspace of R_k
SU_2	Group of 2×2 unitary matrices with determinant unity
$\sigma^1, \sigma^2, \sigma^3$	Pauli spin matrices
$T(x)$	Operator function
Tr	Trace
U_g	Representation of 0_3
u	Element of the group SU_2
v	Element of 0_3
(x, y)	Scalar product
x_1, x_2, x_3	Spatial coordinates
x_4	Time coordinate
$\vec{x}$	Radius vector with components (x_1, x_2, x_3)
Z_k	Spinor representation
$\tilde{z}$	Right coset
ψ, θ, ϕ	Three angles equivalent to the Euler angles
ξ^k	Complex numbers
η	Minkowskian metric
$\sim$	Equivalence relationship
[ ]	Commutator

Page numbers given in italics refer to references cited in the book.